The ASA Press

The ASA Press imprint represents a collaboration between the Acoustical Society of America and Springer dedicated to encouraging the publication of important new books in acoustics. Published titles are intended to reflect the full range of research in acoustics. ASA Press books can include all types of books published by Springer and may appear in any appropriate Springer book series.

Editorial Board

 ASA Press

Acoustical Society of America

The mission of the **Acoustical Society of America** (www.acousticalsociety.org) is to increase and diffuse the knowledge of acoustics and promote its practical applications. The ASA is recognized as the world's premier international scientific society in acoustics, and counts among its more than 7,000 members professionals in the fields of bioacoustics, engineering, architecture, speech, music, oceanography, signal processing, sound and vibration, and noise control.

Since its first meeting in 1929, The Acoustical Society of America has enjoyed a healthy growth in membership and in stature. The present membership of approximately 7,000 includes leaders in acoustics in the United States of America and other countries. The Society has attracted members from various fields related to sound including engineering, physics, oceanography, life sciences, noise and noise control, architectural acoustics; psychological and physiological acoustics; applied acoustics; music and musical instruments; speech communication; ultrasonics, radiation, and scattering; mechanical vibrations and shock; underwater sound; aeroacoustics; macrosonics; acoustical signal processing; bioacoustics; and many more topics.

To assure adequate attention to these separate fields and to new ones that may develop, the Society establishes technical committees and technical groups charged with keeping abreast of developments and needs of the membership in their specialized fields. This diversity and the opportunity it provides for interchange of knowledge and points of view has become one of the strengths of the Society.

The Society's publishing program has historically included the *Journal of the Acoustical Society of America*, the magazine *Acoustics Today* (www.acousticstoday. org), a newsletter, and various books authored by its members across the many topical areas of acoustics. In addition, ASA members are involved in the development of acoustical standards concerned with terminology, measurement procedures, and criteria for determining the effects of noise and vibration.

E.F.F. Chladni

Treatise on Acoustics

The First Comprehensive English
Translation of E.F.F. Chladni's
Traité d'Acoustique

Translated by
Robert T. Beyer

 ASA Press

 Springer

E.F.F. Chladni
Wittenberg, Germany
Paris, France

Translated by
Robert T. Beyer (1920–2008)
Brown University
Providence, RI, USA

ISBN 978-3-319-20360-7   ISBN 978-3-319-20361-4   (eBook)
DOI 10.1007/978-3-319-20361-4

Library of Congress Control Number: 2015945948

Springer Cham Heidelberg New York Dordrecht London

Printed on acid-free paper

Springer International Publishing AG Switzerland is part of Springer Science+Business Media (www.springer.com)

Note on the Translation

This work was translated by Robert T. Beyer, Ph.D. (1920–2008), noted acoustician, Professor of Physics at Brown University, and Gold Medal recipient of the Acoustical Society of America. Along with other projects, Dr. Beyer worked on this translation over the last 10 years of his life. As a labor of love, this project was prepared for publication by his children and grandchildren.

The original text includes eight fold-out plates of all of the figures referenced in the text. Since Chladni does not reference the figures in strict numerical order, we have followed the original, and included them at the end of the text, for reference, in 16 standard pages.

Footnotes from the original manuscript have been retained. Additional footnotes have been added to clarify the translation for the modern reader. The key for authorship of the newly added footnotes is as follows:

RTB — Robert T. Beyer, Ph.D.
MAB — Margaret Anne Beyer
CBH — Catherine Beyer Hurst
RRB — Roberta Rea Beyer
TDR — Thomas D. Rossing, Ph.D.
JPC — James P. Cottingham, Ph.D.
GB — Guillaume Bouchoux, Ph.D.

With special thanks to Margaret Beyer, who began assisting her father with this project in 2003, for spearheading this effort and organizing the manuscript, to Rick Beyer for reproducing the plates, to Roberta Beyer for assistance with the French, and to Catherine Beyer Hurst for copy editing the final manuscript. We would also like to thank the Acoustical Society of America for their support, and especially Ken Foote for his unflagging enthusiasm. And, finally, thanks are also due to Thomas Rossing, Ph.D., who applied his depth of knowledge in musical acoustics to a thorough review of the translation; to Guillaume Bouchoux, Ph.D., a bilingual acoustician, who checked passages in the original French when needed; and to James Cottingham, Ph.D., Chair of The Acoustical Society of America Press Editorial Board, who also reviewed the text, monitored and coordinated changes suggested by the other readers, and answered numerous editorial questions. It really does take a village!

In loving memory of our father and grandfather.

Catherine Beyer Hurst
Margaret Beyer
Rick Beyer
Mary Beyer Trotter
Brian Hurst
Tim Hurst
Roberta Beyer
Andy Beyer
Julie Trotter
Rachel Trotter
Faith Trotter

Contents

Translator's Introduction: Chladni and the State of Acoustics in 1800

[This introduction is adapted and has been modified from the material on Chladni found in the introductory chapters of *Sounds of Our Times: Two Hundred Years of Acoustics* by Robert Beyer. ISBN 978-0-387-98435-3. Reproduced with permission]

In 1802, Ernst F.F. Chladni (1756–1827) published, in German, *Die Akustik*.[1] Chladni himself translated the book into French, a book "in which I have abridged, changed and added a great deal." He published this volume in 1809 as *Traité d'Acoustique*, dedicating it to Napoleon (a wise choice, no doubt, during that "sun of Austerlitz," especially since the French government contributed funds to support the translation and revision). It is the translation of this French language volume that is presented here.[2]

By just glancing through *Traité d'Acoustique*, we can see several features that distinguish the book from a modern one. The first is the almost complete absence of mathematics. Acoustics, as it was studied at the time, at least, in the mind of Chladni, and aside from music and vibrating structures, was largely a science of observations and descriptions. Magnificent mathematics had been developed by Euler, d'Alembert, and Lagrange in the eighteenth century and applied by them to acoustical problems, but Chladni clearly passed over the details of this mathematics in writing his treatise.

---

[1] E.F.F. Chladni, *Die Akustik*, Breitkopf & Hartel, Leipzig, 1802.

[2] E.F.F. Chladni, *Traité d'Acoustique*, Courcier, Paris, 1809.

Ernst F.F. Chladni (1756–1827) (From D. Miller.[3])

A second difference is the emphasis on vibrations. If one had any doubt that vibrations have long been recognized as an integral part of acoustics, a reading of Chladni would eliminate that misconception. Of the four sections of Chladni's book, the one devoted to vibrations comprises more than 60 % of the text. Propagation of sound through air and other gases covers about 15 %, while the remaining 25 % of the volume is divided among propagation in liquids and solids, musical scales, speech, and hearing. (Parenthetically we might note that, even though the period around 1800 was far from a quiet time in the world, there is virtually nothing in the book on noise).

In this book, Chladni divides the subject of acoustics into sources of sound, the passage of sound through matter, and its reception. Perhaps the first problem of sound propagation was the question of whether or not air (or other material) was necessary for sustaining its propagation. By 1800, this was thought to have been long settled. One of the oldest and most frequently repeated experiments in acoustics is the use of a bell or other mechanical source of sound in a chamber that had been evacuated to some extent. This experiment was first carried out by Sagredo in 1615, and repeated a number of times over the next 200 years, with the conclusion, as of 1800, that it had been proven that sound could not travel through a vacuum.

The next question was that of the velocity of propagation in air. By 1800, accurate measurements of the velocity of sound in air had existed for more than 150 years. However, the theoretical basis for calculation of the sound velocity in gases still remained a puzzle. For its calculation, the scientists of the day went back to Newton. Chladni, using Newton's method, cites velocities for a number of gases,

---

[3] D.C. Miller, *Anecdotal History of the Science of Sound*, Macmillan, New York, NY, 1935, opposite pp. 24, 51.

such as oxygen and carbon dioxide, that come close to currently accepted values, but was far off in his estimate of the value in hydrogen—680–810 m/sec as against today's accepted value of 1240 m/sec. This problem was still not entirely settled in 1800. It must be remembered that much of our understanding of the behavior of gases, and of thermodynamics, stems from work done in the first quarter of the nineteenth century. Further progress in our knowledge of the velocity of sound had to wait for such a development.

The effect of the temperature on the velocity of propagation of sound was known qualitatively in 1800, but the precise connection had to await a better knowledge of the thermal properties of gases.

In 1800, there was virtually no knowledge of sound transmission in liquids. The velocity of sound in water had not yet been measured. However, a rather good theoretical value for this velocity is given by Young, who noted that the elasticity (we would call this the bulk modulus) of water had been measured by Canton in 1762 and found to be 22,000 times that of air. Using the elasticity data of Canton, Chladni is able to determine the sound velocity in a number of liquids with accuracy similar to that obtained for water.

A measurement of the velocity of sound in solids did exist in 1800, and is described by Chladni in this book. He compares the musical pitch emanating from a struck solid bar (undergoing longitudinal oscillations) with the pitch of the (standing) wave in a closed, air-filled pipe of the same length. Arguing that the difference is due to the difference in the two sound velocities, he comes up with values of the sound velocity. Chladni's wording is a bit obscure, and his values are too low by nearly 15 %. Nevertheless, he did demonstrate that sound velocity is considerably higher in solids than in gases or liquids.

In his discussion of echo, Chladni notes that it is possible (by having two reflecting surfaces facing one another) to have multiple echoes. He remarks that the ear is capable of distinguishing eight or nine different sounds in a single second. Therefore, when these repeated echoes take place more rapidly than eight or nine per second, he states that the phenomenon is known as resonance. We would define resonance somewhat differently today.

In considering sound intensity, Chladni notes that it depends on: the size of the sonorous body, the intensity of the vibrations of that body, the frequency of the vibrations, the distance at which the sound is heard, the density of the air, the direction in which the sound is heard, and the direction of the wind. We would recognize today that these dependences are a mixture of what we call the intrinsic intensity ($\frac{1}{2}\rho c v_0^2$, where ρ is the local gas density, c the speed of sound, and v_0 the amplitude of the displacement velocity), and such related quantities as the strength of the source, the directionality of the source, and the attenuation due to geometric spreading and sound absorption. All these quantities were not well sorted out in 1800.

In 1800, the available means for the production of sound were the human voice, musical instruments, cannons and other explosive devices, and natural phenomena such as animal sounds and thunder. It is not surprising, therefore, that Chladni (and others of the time) used music as the basis on which to build almost all of acoustics. When dealing with vibrating strings, they were concerned with stringed musical instruments. Vibrating air columns were of interest because of organ pipes, and also various musical horns, while stretched membranes were related to drums. Almost every subject in Chladni's text is studied from the point of view of music. The great advances of theoretical acoustics in the eighteenth century were perhaps due to the common interests of music patrons, researchers, and the listening audiences.

Well before 1800, the understanding of music led to two great contributions to the science of sound. First, it emphasized the importance of ratios for different tones. The simple ratios appropriate for all the notes on the diatonic scale were known, and musicians with trained ears could easily identify the pitches of the various notes, starting from some accepted standard, such as middle C or, more commonly, the A above middle C. At the same time, it was also known that the pitch of a musical note was measured by its frequency of oscillation.

Because of earlier observations by Mersenne and Sauveur, Chladni is able to base his discussion of musical scales on the grounds of a knowledge of the frequencies involved in the tones of the scale, even though there remained some uncertainty as to what was the appropriate standard for middle C. One must note, however, that Chladni devotes most of his attention, in musical matters, to discussing ratios of frequencies of the different tones, where the quantities are far more accurate, rather than the absolute values of these frequencies. Chladni was aware that the human ear could hear tones as low as 30 Hz and up to 8000–12,000 Hz—which goes considerably beyond the range of frequencies achieved by most musical instruments.

Somewhere between the production and detection of sound is Chladni's own work on vibrating plates, since the study involved not only the production of sound by the vibrating plates, but also the experimental technique of identifying the vibrations. Chladni was well aware of the vibration of strings, and of the localization of nodal points in a standing wave, and the theoretical work of Lagrange and others gave a strong foundation to the subject. There was, however, no theory of vibrating plates.

Chladni was drawn to a study of the vibrations of plates from work done by Lichtenberg who scattered "electrified powder over an electrified resin-cake, the arrangement of the powder revealing the electric condition of the surface."[4] In Chladni's first work, reported in 1787, he held fixed one or more points on a plate and stroked the side of the plate with the bow of a violin. In order to render the effect of the vibrations visible, he placed a little sand on the plate. The sand "was thrown aside by the trembling of the vibrating parts (of the plate) and accumulated on the nodal lines." Chladni must have been fascinated by the patterns taken on by the sand particles, a fascination that these "Chladni figures" continue to generate. In those days before photography, he included hundreds of drawings of different

_____

[4] These "Lichtenberg figures" are what Chladni refers to as "electric figures" in the preface to *Traité d'Acoustique.—JPC*

modes of excitation for triangular, circular, square, and even elliptical plates, in this book.

The only instrument available for sound reception in 1800 was the ear. By the time of Chladni, the structures of the exterior and middle portions of the ear were quite well understood. Chladni recognized that the impulses of sound received in the outer ear were transmitted through the small bones (ossicles) of the middle ear, more or less faithfully, to the cochlea. While the gross features of the cochlea were well described by him, its role, so far as he was concerned, was very much that of a "black box." The sound impulses impinged on one end, and, somehow or other, the sensations were picked up by the auditory nerve at the other and transmitted to the brain. Chladni offered the (incorrect) opinion that the signals from the ossicles affected the cochlea as a whole rather than locally. Further developments in this area had to wait another century.

A phenomenon was observed in the eighteenth century that was later to have important consequences in physical acoustics, although this did not occur until much later. This is the phenomenon that was known as Tartini tones: When two musical sounds of different pitch are sounded simultaneously and loudly, a tone is heard with a pitch equal to the difference in the pitches of the two tones. Chladni and others studied the problem and concluded that it was a form of beats. Under the usual description of beats, when the difference in pitch is small (5–10 beats per second), we can distinguish them clearly. When the number of beats per second increases, we first distinguish the unpleasant sounds of dissonance, but as the number gets very large, they argued, we ultimately hear the pure tone of the difference frequency. While this interpretation was later proved to be incorrect, it satisfied the acoustics community for the next half-century.

In looking back over the period before the publication of *Traité d'Acoustique*, one is humbled by the accomplishments of acousticians who worked with an almost complete lack of anything we would call apparatus. What they had were the human voice and ear, musical instruments, bells and tuning forks, vibrating strings and plates, the basic equations developed over the years, and a great deal of ingenuity and resourcefulness. In a day when we can scarcely add numbers without a calculator, or perform an experiment without vast arrays of electronic equipment, we can only marvel at the success of their ingenuity and resourcefulness.

Robert T. Beyer

TREATISE
on
ACOUSTICS
(TRAITÉ D'ACOUSTIQUE)

by

E. F. F. CHLADNI

Doctor of Philosophy and of Law; Member of the Royal Society of Haarlem, of the Society of Friends of Natural Science of Berlin, of the Academy of Applied Science of Erfort, and of the Departmental Society of Mayence; Correspondent of the Imperial Academy of Saint Petersburg, the Royal Societies of Göttingen and Munich, of the Philomatic Society of Paris, and of the Batavian Society of Rotterdam.

PARIS – 1809

Napoleon the Great

Has Deigned to Agree

To the Dedication of this Work

After Viewing

The Fundamental Experiments

Preface

While the other branches of physics have been moving forward, acoustics has always lagged behind. The sounding vibrations of most elastic bodies were entirely unknown, and, ordinarily, only the transverse vibrations of a string have been considered. These have therefore been regarded as the basis of all harmony and we have tried to extend its laws to all other sounding bodies. The results of the research that many scholars such as Daniel Bernoulli, Euler, Lagrange, Giordano Riccati, and others conducted on various acoustical subjects were not introduced into the mass of knowledge that is expounded in treatises on physics. This is what caused me to undertake the development of this vast uncultivated field, and to uncover the laws of vibrations and their different modifications in all sorts of sounding bodies, according to the research of the principal geometricians and physicists, and according to numerous experiments. My aim was to give the results of both theory and experiment in a sufficiently clear and precise manner, so that at least the greatest part would be within the range of those who had only a slight acquaintance with physics and mathematics. Those who are more advanced will lose nothing because they will find sufficient data for further research and because I have always cited the principal works and dissertations which will serve for their further instruction.

Among the experiments whose results I have related here, there are none that I have not performed myself, and that I cannot repeat. I respect nature too much to want to attribute anything to it that would perhaps be only a play of my imagination.

Everything that is printed at the end of several paragraphs by way of the strictest justification contains notes or additions to the text of the paragraph, marked in such a way that the connection of the materials should suffer the least interruption.[5]

In publishing this *Treatise on Acoustics*, I am responding to the desires of several persons whose commendations and kindnesses provide a powerful motive for me to attempt to merit them by such useful work in science. I am especially honored by

---

[5] In this edition, these notes appear in a greyed-in box.—*MAB*

the fact that I have been brought to this purpose by the illustrious author of *Celestial Mechanics*, who is also respected for his benevolence and his zeal in encouraging those who work in the sciences, as well as by what he has done for the increase in human knowledge. My German work could not be translated into another language without some appropriate changes; another translator would perhaps have problems of his own, and he might have had me saying something that I did not say.[6]

Therefore, I have undertaken this work myself. It must therefore be excused if a stranger who has not spent a long time in France does not always express himself with sufficient clarity. Friends who have had the kindness to preview my work have corrected some of the mistakes; I do not know if outside critics will be as indulgent as my friends; however, if they wish to be fair, they will spend more time on the subject than on the language. It has often been asked, by what chance did I happen to make certain discoveries. But chance has never favored me; to be successful I had, nearly always, to employ opinionated perseverance. Following the advice of several estimable persons, I must add here some remarks concerning the story of my discoveries, these being the result of individual circumstances. I believe that these circumstances should interest a number of readers.

My father (First Professor of Law at Wittenberg in Saxony, one of the most esteemed judicial consultants in this county because of his activity, his talent, and his probity) had provided a good instruction for me, but it was at home, and then in the provincial school of Grimme; my education therefore allowed me little liberty, so that, whereas others may recall their youth as the happiest period of their lives, I cannot say the same about mine. This continual constraint, which was not at all necessary, because I was not disposed to abuse my freedom, produced a completely contrary effect on me, imprinting an almost irresistible desire to choose my occupations myself, to travel, to struggle against adverse circumstances, and so on. Returning to Wittenberg, it was necessary, following the will of my father, for me to apply myself to jurisprudence.

After having studies at Wittenberg and at Leipzig, and having satisfied the requirements, I obtained the degree of Doctor in Philosophy and in Law at Leipzig. Fate seemed to have destined me to remain always at Wittenberg and for me to obtain employment as a Professor of Law. But, after the death of my father, I quit jurisprudence, because it conformed too little to my tastes, and applied myself principally to the study of nature, which had always been my secondary occupation, and that dearest to me. As an amateur musician, of which I had begun to learn the basic elements (a little late), in my nineteenth year, I noted that the theory of sound was the most neglected part of physics, which gave birth in me the desire of remedying this defect and of being of use to this part of physics by making a number of discoveries. In performing numerous very imperfect experiments in 1785, I had observed that a plate of glass or of metal gave off different tones when it was clamped and struck in different places, but I did not find any instruction

---

[6] These rules have been violated in the present translation, but then Chladni never undertook the translation into English. Sorry Ernst!—*RTB*

as to the nature of these types of vibrations. At that time, journals had given some publicity to a musical instrument, made in Italy by the Abbé Mazzochi.

This instrument consisted of glass bells to which he had applied one or two bowings of a violin; from this I conceived the idea of using the bow of a violin to examine the vibrations of different vibrating bodies. When I applied the bow to a round brass plate, stationary at its center, the plate made different sounds which, when compared with one another, were proportional to the squares of 2, 3, 4, 5, etc.; but the nature of the movements to which these tones correspond, and the means of producing each of these movements at will were still unknown to me. Experiments on the electric figures that are formed on a plate of dusted resin, discovered and published by Lichtenberg (in the *Memoirs of the Royal Society of Göttingen*), led me to presume that the different motions of a vibrating point ought also to show different appearances if one were to put a little sand or other similar material on its surface.[7] In making use of such means, the first figure that I observed, on the round plate of which I spoke, resembled a star with 10 or 12 rays, a little like Fig. 102a, and the most acute sound was, in the series cited, the one which corresponded to the square of the number of diametral lines. One can judge my astonishment in viewing this phenomenon, which no one had previously seen. After having reflected on the nature of these motions, I did not find it difficult to vary and to multiply these experiments, from which results followed rather rapidly. My first dissertation, which contained some research on the vibrations of a round plate, of a square plate, of a ring, of a bell, etc., came out in Leipzig in 1787. The results of the research that I have carried out since that time on longitudinal vibrations and on other subjects of acoustics have appeared in several German journals and in the papers of various Societies.

Finally, after having made still more experiments and after having consulted further the research that others have performed, I joined the results together, so far as it was possible, in my *Treatise on Acoustics*, which came out in Leipzig in German [*Die Akustik*]. The present work is a translation of that book, a work that I have abridged, changed, and expanded a great deal, according to what seemed to me to be most appropriate.

It was Lichtenberg, whose ingenious ideas in physics I found most interesting, who gave a second impulse to the advance of my ideas. Being in Göttingen in 1792, I asked his opinion on the nature of fiery meteors (which are called *bolides* or fireballs), whose phenomena, such as a flame, smoke, and explosions, were little in conformity with the electrical phenomena with which they are identified. He responded that the best way of explaining these phenomena would be to attribute to these meteors an origin that is *cosmic* rather than *terrestrial*, that is to say, to suppose that it was something foreign that had arrived from the outside into our atmosphere, a little as Seneca had well explained the nature of comets, which,

---

[7] In 1777 Lichtenberg obtained what are now called Lichtenberg figures by discharging a high voltage point to the surface of an insulator and recording the resulting radial patterns by sprinkling various powdered materials onto the surface.—*JPC*

nevertheless, had a long time after been regarded as atmospheric meteors until the Saxon pastor, Dörfel, had shown that Seneca was correct. Struck by this observation of Lichtenberg, I consulted works and dissertations that contained reports of similar meteors, and of stones or masses of iron that one had seen fall at one time or another as the result of a similar meteor, and I finally published my research in a dissertation that came out in Leipzig in 1794, the French translation of which, made by Eugène Coquebert-Mombret, is to be found in Vol. V of the *Journal des Mines*. I demonstrated the following observations in this paper (before the fall of rocks [meteorites] that took place at Siena on June 15, 1794).[8] First, that the reports that had been given concerning rocks and masses of iron that had fallen with great impact as the result of a meteor fireball were not fictions or illusions, but observations of a real phenomenon; second, that these masses and these meteors were something foreign to our globe, and had arrived from the outside. In the beginning, no one agreed with me; several German critics even supposed that I had not advanced this idea seriously, but with a somewhat evil intent of seeing what fraction of physicists accepted them and how far the credulity of these individuals might extend.

In France, Mr. Pictet was the first to call to the attention of physicists as to what my paper contained (Vol. 16 of the British Library); but no one believed in the possibility of a fall of rocks until Howard's dissertation in 1802. And in 1803, the fall of rocks took place at L'Aigle, and it was established by Mr. Biot that I was not given to fits of imagination; this finding has been further demonstrated by numerous meteors which have been observed, and by the research that has been conducted. In the *Bulletin of the Philomatic Society* of April, 1809, I published a catalogue of sizeable meteors that have been observed to date, to which one could add still others that have been found; including those which Soldani has cited in Vol. 9 of the Memoirs of the Academy at Siena.

The invention of the *euphone* and the *clavicylinder*, and their execution under less than favorable circumstances, cost me a great deal of time, more work, and more expense than my research on the nature of sound, of which these two instruments are practical applications. Those who have worked in a similar area, as, for example, those who have tried to perfect the glass harmonica,[9] know how many unforeseen difficulties are encountered in such work. Too often, when one wishes to put into practice those ideas which appear to conform to theory, nature, being consulted by means of experiments and tests, denies our conjectures and places insurmountable obstacles in front of us, obstacles that had not been foreseen. Then, after having worked without success for a long period of time, it becomes

---

[8] There was an eruption of Mt. Vesuvius on June 15, 1794 near Siena, Italy, followed by the fall of a meteorite the following day, 200 km to the north.—*RTB*

[9] There are two types of glass harmonica: glass harps, which consist of wineglasses rubbed by the fingers, and the armonica, invented by Benjamin Franklin, in which glass bowls are mounted on a shaft so that they rotate (partially in water) as the shaft rotates. The first type was probably most familiar to Chladni. *When the word "harmonica" is used throughout this text, it is understood to mean "glass harmonica," not the modern wind instrument.*—*TDR*

necessary to set aside everything that has been done and begin anew. But the least success makes us forget all these tests of our patience.

The *euphone*, invented in 1789 and perfected in 1790, consists exteriorly of small cylinders of glass that one rubs longitudinally with fingers moistened by water. These cylinders, of the thickness of a writing pen, are all of equal length, and the difference in the tone is produced by an interior mechanism. The tone resembles more that of the harmonica than that of any other instrument. In several trips that I have made to Germany, to St. Petersburg, and to Copenhagen, this instrument has received widespread approval. The one that I had with me most recently has been destroyed during the voyage by some accident, but the construction of another, on which I am now working, is almost complete.

The *clavicylinder*, invented at the beginning of 1800 and perfected since that time, consists of a keyboard, and behind this keyboard a glass cylinder, which is turned by means of a pedal and a leaden wheel. This cylinder is not itself the sounding body, as in the bells of the harmonica, but produces the tone by its rubbing of the interior mechanism. The principal quality of this instrument is its ability to prolong the sound at will with all the nuances of crescendo and diminuendo, as one augments or diminishes the pressure on the touch. The instrument is never out of tune. The reports of the Institut de France and of the Imperial Conservatory of Music, which have given very favorable judgment on this instrument, relieve me of the necessity of talking of it at length.

For some time, I have again been occupied with conducting research and experiments on different methods of constructing a *euphone* or *clavicylinder*. As the possible methods are very numerous, and as it is difficult to judge in advance the preference for one or the other, this subject will continue to keep me busy. Each invention being the property of its author, I do not merit reproach for the fact that I have still not published the internal mechanism and the construction. I have still not lost the hope that there will come a time when a compensation proportional to the sacrifices I have made will allow me to publish everything concerning the theory and construction of these instruments.

The great value that I attach to the approval of the Institut de France, and the respect that I have toward that leader of scholarly societies, whose members I have the honor of knowing personally, has encouraged me to have printed at the end of this work the prize program that it has proposed on the mathematical theory of vibrating plates and the reports that it has kindly wished to make on my research and their applications to the arts.

E.F.F. Chladni

Preliminary Observations

1. Explanation of the Words *Sound* and *Noise*

Every possible motion is either progressive, rotary, or vibratory. A sufficiently rapid vibratory motion that is strong enough to affect the hearing organ is a *sound*.

If the vibrations of a sounding body are distinguishable, both in their frequency[1] and in their change in shape, they are called *distinct sound* or *sound* properly called, in order to distinguish them from *noise*, or indistinguishable vibrations.

Elasticity is the actuating force for sound. A sounding body can be elastic either by *tension*, by *compression*, or by *shear stiffness*. To perceive a sound, it is necessary that there exists a continuation of some kind of matter between the sounding body and the organs of hearing. The air is the ordinary conductor of sound, but all liquids and solid bodies propagate sound with greater speed and force than does the air. It is therefore quite out of place, in treatises on physics, to assign the theory of sound to the theory of gases; rather, one must make it a part of the theory of motion and join it to the theory of pendulums.

---

[1] Chladni uses the term *vitesse* here, which could also mean velocity. Since he is referring to the rate at which the signal changes its shape, it is appropriately translated as frequency.—*RTB*

© Springer International Publishing Switzerland 2015
E.F.F. Chladni, R.T. Beyer, *Treatise on Acoustics*,
DOI 10.1007/978-3-319-20361-4_1

2. Subjects of Acoustics

Acoustics is the theory of sound. The subjects of acoustics are:

1. The numerical ratios of the vibrations
2. The characteristic vibrations of sounding bodies
3. Transmitted vibrations, or the propagation of sound
4. The sensation of sound, or hearing

Number 1 forms the arithmetic part, numbers 2 and 3 the mechanical part, and number 4 the physiological part of acoustics.

Part I
Numerical Ratios of Vibrations

Section 1: Primary Ratios

3. Grave and Acute Tones

The French word *son* (sound), as used here, expresses the pitch of the vibrations. An *acute sound* differs from a *grave sound* by the greater (acute) or smaller (grave) number of vibrations that it executes in a given interval of time.

> The word sound, therefore, has three very different meanings: it expresses first, everything that can be heard (in German, *Schall*); second, regular or perceptible vibrations (in German, *Klang*) as opposed to noise; third, the frequency of the vibrations (in German, *Ton*). The word *ton* is not used in French, as it is in other languages, to express the frequency of the vibrations in each sound. This word has several other meanings; for example, it expresses a major second. It also indicates the mode, the intensity with which one executes the music, etc.

4. Explanation of the Words *Interval, Melody, Chord, Harmony*, etc.

An *interval* is the ratio of the frequency of one sound to another. Usually, a very deep or grave sound is regarded as the basis for comparison with a more acute sound.

A *melody* is a sequence of tones.

A *chord* is the coexistence of several tones.

© Springer International Publishing Switzerland 2015
E.F.F. Chladni, R.T. Beyer, *Treatise on Acoustics,*
DOI 10.1007/978-3-319-20361-4_2

Harmony is a sequence of chords, or the coexistence of several melodies.

Music makes use of materials, for which acoustics furnishes the theory, in order to excite sensations.

5. Absolute Frequency of the Vibrations of Each Tone

(*Note*: *Description of a tonometer*.) In the deepest (gravest) sounds that are perceptible to the human ear, the vibrating body makes at least 30 vibrations per second; but acute sounds can be sensed in which up to 8000 or 12,000 vibrations occur per second. We do not go far from the truth if, to facilitate comparison of the absolute number of vibrations to relative numbers, we regard each *do*[1] as representing the power of two, taking the fundamental *do* to be unity. We therefore attribute to the lowest *do* on the piano or violoncello the value of 128 vibrations (or simple oscillations per second), which is enough to find the absolute number of vibrations of every other sound, by multiplying the relative numbers (Pars. 19 and 26) for the first lower octave by 128, for the second by 256, for the first octave above by 512, etc.

> The value that one has assigned to these instruments has not always been the same in different countries and in different eras. Thus, Euler assigned a value of 118 to *do* (in his *Tentam. nov. theor. mus.*, ch. 1) and 125 in another (in *Nov. Comment. Acad. Petrop.* vol. 16). Marpurg at Berlin found this same latter result while Sarti (*Nov. Act. Acad. Petrop.* 1796) noted that the *la* of the third string of a violin makes 436 double, or 872 single, vibrations per second, in St. Petersburg.[2] This gives a value of approximately 131 vibrations per second for this same *do*. There has been a tendency to increase these values in tuning instruments since the days of Euler and Marpurg, and, at present, in several orchestras, they have gone somewhat beyond the number of vibrations that result from using powers of two. However, one can still adopt some power of two for the average value of *do*.[3]

(continued)

---

[1] Chladni used the symbol *ut* for the first note in the musical scale, but we have replaced it by *do* regularly used in modern texts. We have left the seventh note as *si*, since both *si* and *ti* are used today in different countries.—*RTB*

[2] In the early mathematical studies of music, a debate existed between those who used the number of half oscillations (simple vibrations) per second, and those who used the number of full oscillations (double vibrations). The latter school won out.—*RTB*

[3] This "powers-of-two" scale is no longer used in music but is commonly employed in instruction in physics, and is sometimes called the physicists' scale.—*RTB*

I have found a very simple method of determining, by the judgment of the eyes and ears, the number of vibrations that corresponds to each sound. It is necessary that a vibrating body be of a sufficient length (which can be shortened if required) for one to see and count the vibrations that take place in a certain interval of time, in order to compare the sounds and the lengths of the parts with the number of vibrations counted, and also compare them to the length of the entire body. I used very long strings for this effect; but I did not succeed, because the vibrations of the different parts were added to vibrations of the entire string, as well as many circular and lateral motions, which made the accuracy of the observations difficult. It is therefore preferable to make use of a thin sheet of iron or brass, of about a half-line of thickness and of a half inch in width and of a length sufficient to vibrate very slowly. It was necessary that the thickness be exactly uniform. A metal wire could have been used, but the width of the sheet served to prevent lateral motions. The frequency of vibrations of such a sheet is inversely proportional to the squares of their lengths, when other conditions remain the same. The end of the sheet was clamped in a fixed vise, making it project out, more or less, to the length at which it makes, in each second of time, a certain number of vibrations visible to the eye. These vibrations can be compared to the oscillations of a second pendulum, which is understandable, as it is done in astronomical observations. When one is a little used to it, it is not difficult to count up to eight vibrations per second. I propose to use 4 vibrations per second to mark, exactly, the length of the projection from the sheet and to divide it into two, four, eight, and other numbers of parts. If one fixes the sheet in the vise in such a way that one half is extended, it will make 16 vibrations per second. These vibrations will be too slow to be heard and too rapid to be counted; but one can hear a distinct sound by making the sheet vibrate in two unequal portions, so as to establish a node of vibrations, at a distance from the free end, slightly less than one third of its length. This sound, which I call the *second sound*,[4] because it corresponds to the second figure of the plate (Fig. 21), will make 100 vibrations per second, like the stationary sound of Sauveur. It will be *sol#*, approximately a major third below *do*, the lowest note on the keyboard. If the part sticking out from the vise is shortened, so that it equals one fourth of the thickness of the plate, it will make 64 vibrations, one octave below the first octave on the keyboard. The second sound, which makes 400 vibrations, will give the *sol*, two octaves higher than the one which has 100 vibrations. Whatever may be the manner in which it partakes

(continued)

---

[4] What Chladni refers to is the second partial (often the second harmonic).—*TDR*

of the motion of the plate, the results of these experiments conform very well to the theory. The best manner of producing this sound will be to use the bow of a violin. Before making the experiments, one should read what I have said regarding the transverse vibrations of a straight rod, in Sect. 5 of Part II.

6. Difference of Consonant and Dissonant Intervals

The greater or lesser simplicity of the numerical ratios of the vibrations is the sole basis of harmony. An *interval* is *consonant* when the ratio is very simple; when the ratio is less simple, the interval is *dissonant*. The consonant intervals can be expressed by the numbers 1, 2, 3, 4, 5, 6 or by 1, 3, 5 and the doubling of any of these numbers; dissonance results from different combinations of the same numbers. A consonant interval is pleasing by itself, while dissonance is only pleasing when it is returned and when it passes to another simpler ratio.

The ear, without counting the numbers themselves, perceives the effect of the relationships of the concurrence of simultaneous vibrations when they arrive together. It does for time what the eye does for space, when it is affected in an agreeable manner by the fair relationship of forms, without measuring and without calculating the ratios themselves.

Leibniz expressed himself very well on this subject (*Epistolae ad diversos*, vol. 1, epist. 154):

Musica est exercitium arithmeticae occultem nescient is se numerare animi; multa enim facit in perceptionibis confuies seu insensibilibus, quae distincta aperceptione notare nequit. Errant enim, qui nihil in animá fieri putant, ujus ipsa non sit conscia. Anima igitur etsi se nunerare non sentiat, sentit tamen hujus numerationis insensibilis effectum, seu voluptatem in consonantiis, molestiam in dissonantiis inde resultantem. Ex multis enim congruentiis insensibilibus oritur voluptas, etc.[5]

Descartes also proposed the same principles (*Tract de homine*, p. 3, sect. 36, and *Comp. mus.*).

It does not conform to nature to want to derive, as several authors do, all the harmony of vibrations, and in particular the coexistence of some with the

(continued)

---

[5] Unlike Chladni, we do not assume that the modern reader readily understands Latin. The first line of this quote from Leibniz is a famous definition of music: "Music is a hidden arithmetic exercise of a soul that does not know it is counting." Leibniz says that the pleasure we obtain from music comes from this unconscious counting.—*CBH*

fundamental sound, from those of a string. A string is only one kind of sounding body. In many other bodies, the general laws of vibrations, which one had not known, are modified differently; consequently, one cannot apply the laws of a sounding body to that which must be common to all. A monochord cannot serve to establish the principles of harmony; but only to give an idea of the effect of ratios.

7. Unison and the Octave

The simplest ratio is 1:1, in which two vibratory motions occur at the same time, and is known as *unison*.

The interval 1:2, in which the frequency of one vibration is double that of the other, is known as an *octave*. It is called thus because it is the eighth step in the ordinary scale, as each other interval takes its name from the step of the scale on which it is found. Experience shows that two sounds which are in the ratio of 1:2 have such a resemblance that we can regard one as the repetition of the other, from whence it follows that:

1. The nature of an interval does not change if one takes the sound of which it is composed one or two octaves lower or higher; that which returns to take double or one half of the smallest number; except that in the case in which one of these numbers becomes larger than the other; for one must regard this interval as an *inversion* of the first. Thus, 2:3, 1:3, 1:6 are the same interval; but 3:4 or 4:3 will be the inverse of that interval.
2. One can regard all the intervals as comprised in a single octave, so that one can express all of them by fractions contained between 1 and 2.

The calculations of intervals are the same as those of fractions.

8. Other Consonant Intervals

All the consonant intervals that can be expressed by the numbers 1, 2, 3, 4, 5, 6, or by doubling of these numbers, when one arranges them between 1 and 2 according to their distance from unity, will be

$$\frac{6}{5}, \quad \frac{5}{4}, \quad \frac{4}{3}, \quad \frac{3}{2}, \quad \frac{8}{5}, \quad \frac{5}{3}$$

of which the last three are also the inversions of the first three. Of all the intervals, that of 3:2, of the *fifth*, is the simplest that the ear perceives as the most perfect consonant after the octave. The *fourth* $\frac{4}{3}$ is an inversion of the fifth. It is consonant by itself, but it is customary in practice to treat it as dissonance because the combinations require a resolution in another interval.[6] The interval $\frac{5}{4}$ is the *major third*, the interval $\frac{6}{5}$ the *minor third*, the *minor sixth* $\frac{8}{5}$, and the *major sixth* $\frac{5}{3}$ are their inversions. Ordinarily, the unison, the octave, and the fifth are called the *perfect consonants*, and the thirds and the sixths, the *imperfect consonants*.

9. Consonant Chords

According to these six consonant intervals, one can judge very easily how many here will be of *chords* or of combinations of more than two *consonant* sounds.

Let $m = 1$, $n = \frac{6}{5}$, $p = \frac{5}{4}$, $q = \frac{4}{3}$, $r = \frac{3}{2}$, $s = \frac{8}{5}$, $t = \frac{5}{3}$. The possible combinations will be:

>   *mnp, mpq, mqr, mrs, mst*
>   *mnq, mpr, mqs, mrt*
>   *mnr, mps, mqt*
>   *mns, mpt*
>   *mnt*

In many of these combinations, the last two intervals are not consonant with one another. They are related in *mnp* as $1 : \frac{25}{24}$, in *mnq* as $1 : \frac{10}{9}$, in *mnt* as $1 : \frac{15}{18}$, in *mpq* as $1 : \frac{16}{15}$, in *mps* as $1 : \frac{32}{15}$, in *mqr* as $1 : \frac{9}{8}$, in *mrs* as $1 : \frac{16}{15}$, in *mrt* as $1 : \frac{10}{9}$, and in *mst* as $1 : \frac{25}{24}$. All these combinations do not give a consonant chord. But *mpr* or $1 : \frac{5}{4} : \frac{3}{2}$ makes another, since $\frac{5}{4}$ is to $\frac{3}{2}$ as $1 : \frac{6}{5}$ and *mnt* where $1 : \frac{6}{5} : \frac{3}{2}$ gives another because $\frac{6}{5}$ is as $\frac{3}{2}$ to $1 : \frac{5}{4}$. The combinations *mns, mpt, mqt,* and *mqs* reduce to these two chords if one multiplies or divides the number by 2, and if one expresses them in the smallest numbers. It will never be possible to add a fourth consonant interval to all the others. There will therefore never be a consonant chord composed of more than three tones, except if one wishes to add the octave of one of the three sounds. Such a chord as $1 : \frac{5}{4} : \frac{3}{2}$ or $1 : \frac{6}{5} : \frac{3}{2}$ is a *perfect chord*; the first is the *major perfect chord*, the other the *minor perfect chord*. The consonant combinations $1 : \frac{5}{4} : \frac{5}{3}$ and $1 : \frac{6}{5} : \frac{8}{5}$ (the *chord of the sixth*) and $1 : \frac{4}{3} : \frac{5}{3}$ and $1 : \frac{4}{3} : \frac{8}{5}$ (the *chord of the sixth fourth*) are the inversions of the major perfect chord and the minor perfect chord.

Experience shows that the two perfect chords have different effects. The major is more suitable for expressing joy. It soothes the ear more than the minor. The cause of this different effect is the greater simplicity of the major chord. In reducing these ratios to their lowest terms, the vibrations of the major perfect chord will be as 4:5:6

---

[6] Most musicians now consider the perfect fourth to be consonant.—*TDR*

and those of the minor as $10:12:15$. Both are composed of a major third and a minor, which together make a fifth; the only difference lies on the position of the thirds.

> The manner in which I have shown here the formation of the perfect chords is fundamentally the same as that used by Mr. Mercadier de Belesta (*Système de Musique*, Paris 1776), who set forth several subjects pertaining to the numerical theory of sound better than many others.

10. Dissonant Chords

A *dissonant chord* is one that contains one or more dissonant intervals. The principal of these chords is the *seventh chord*, in which one adds a seventh to a perfect chord. It is susceptible to three inversions, in which one must always regard the sound, which by origin is a seventh, as the dissonance in the position it is found. Several other dissonances result from the delay or anticipation of a sound.

11. Ordinary Scale

The major perfect chord, because of its simplicity, could serve better than the other to find the ordinary *scale* of sound; that is, the series of the most agreeable and suitable of sounds, by which one can pass from a fundamental tone to its octave, and from one octave to another, without losing the sensation of the fundamental sound. The perfect chord of a fundamental tone, joined to its octave, excites the most perfect of the sounds, reinforced by the most suitable consonants. When we regard *do* as the fundamental, we will have:

$$1 \quad : \quad \frac{5}{4} \quad : \quad \frac{3}{2} \quad : \quad 2$$
$$do \qquad mi \qquad sol \qquad do$$

But this is still not a scale because the distances are too great and too unequal. It is therefore necessary to add the perfect chords of these tones, which approach the fundamental more than the others, such as the fifth ($\frac{3}{2}$) and the fourth ($\frac{4}{3}$). The fifth $\frac{3}{2}$ produces, by its perfect chord, the tones $\frac{3}{2} \times \frac{5}{4}$ or $\frac{15}{8} = si$ and $\frac{3}{2} \times \frac{3}{2}$ or $\frac{9}{8} = re.$[7] The fourth $\frac{3}{4} = fa$, which is inserted by itself, produces its major third $\frac{4}{3} \times \frac{5}{4}$ or $\frac{5}{3} = la$.

---

[7] Although $\frac{3}{2} \times \frac{3}{2}$ does not appear to equal $\frac{9}{8}$, this is the original text in both French and German versions of this work.—*MAB*

Its fifth is the same as the octave of the fundamental tone. We will therefore have the scale:

$$1, \quad \frac{9}{8}, \quad \frac{5}{4}, \quad \frac{4}{3}, \quad \frac{3}{2}, \quad \frac{5}{3}, \quad \frac{15}{8}, \quad 2$$

$$do, \quad re, \quad mi, \quad fa, \quad sol, \quad la, \quad si, \quad do$$

This scale has seven intervals of different sizes. The interval of the third to the fourth and that of the seventh to the eighth are approximately one half of the others. One calls the greater intervals *tones* and the smaller ones, *semi-tones*. Each interval draws its enumeration from the interval that it represents, so that the distance from *do* to *re* is a second, from *do* to *mi* a third, from *do* to *fa* a fourth, from *do* to *sol* a fifth, from *do* to *la* a sixth, from *do* to *si* a seventh, and from *do* to *do* an *octave*. If one compares all these tones to the octave above, one will have intervals that must be regarded as the inversions of the previous ones, and which do not differ much from it, so far as the effect and the manner of treating them are concerned. Thus, the distance from *re* to *do* will be a seventh, from *mi* to *do* a sixth, from *fa* to *do* a fifth, from *sol* to *do* a fourth, from *la* to *do* a third, and from *si* to *do* a *second*.

12. Intervals

This scale will acquaint us with most of the dissonant intervals. The first interval is to the second as 1 to $\frac{9}{8}$, or as 8:9 and the second to the third, as $\frac{9}{8}$ to $\frac{5}{4}$ or as 9:10. These two intervals, which differ by $\frac{81}{80}$, we call a tone; the one a *major* tone, the other a *minor* tone.

The major third $1 : \frac{5}{4}$ or 4:5 differs from the minor third $\frac{6}{5}$ by the interval $\frac{25}{24}$, which is the smallest interval one can make practical use of. If an interval is raised or lowered (a *sharp* or a *flat*) to the same degree, the difference is always $\frac{25}{24}$. Each smaller difference is a *comma*. The inversion of a *minor semi-tone* $\frac{25}{24}$ is the *diminished octave* $\frac{24}{25}$.

The difference between the third sound $\frac{5}{4}$ and the fourth $\frac{4}{3}$ is $\frac{16}{15}$; this interval is called a *major semi-tone*; it differs from the minor semi-tone by $\frac{128}{125}$. Sound inversion gives the *major seventh* $\frac{15}{8}$.

The fourth sound differs from the fifth by $\frac{9}{8}$, or a major tone; this one differs from the sixth by $\frac{10}{9}$, or a minor tone. The difference of the major sixth and the minor is that of the third $\frac{5}{24}$. The seventh differs from the octave by $\frac{16}{15}$, or a major semi-tone.

The ratio of these tones gives us still other intervals. The one from *re* to *fa* or $\frac{9}{8} : \frac{4}{3} = \frac{32}{37}$ is a minor third, diminished by a comma $\frac{81}{80}$. That from *fa* to *si* $\frac{4}{3}$ to $\frac{15}{8}$ is an *augmented fourth*, which is also called a *tri-tone* because it results from the combination of three tones; its inversion is the *diminished fifth* $\frac{64}{45}$.

13. Some Other Intervals

There are therefore major and minor seconds, thirds, sixths, and sevenths, but there are no such fifths and fourths. If a fifth or a fourth, as also a major second third, sixth, or seventh, is augmented by a minor semi-tone $\frac{25}{24}$, it is called *augmented*; if a fifth or a fourth as also a minor second, third, sixth, or seventh is lowered by the same interval, it is called *diminished*.

The inversion of a major interval gives a minor and that of a minor gives a major; the inversion of a diminished interval yields an augmented one and the inversion of an augmented one gives a diminished. The augmented and diminished intervals of which we will make use of include:

The *augmented second* $\frac{9}{8} \times \frac{25}{24} = \frac{75}{64}$ or $\frac{10}{9} \times \frac{25}{24} = \frac{125}{108}$, and its inversion, the *diminished*
   seventh $\frac{16}{9} \times \frac{24}{25} = \frac{128}{75}$ or $\frac{9}{5} \times \frac{24}{25} = \frac{216}{125}$.
The *diminished third* $\frac{6}{5} \times \frac{24}{25} = \frac{144}{125}$ and its inversion, the *augmented sixth* $\frac{5}{3} \times \frac{25}{24} = \frac{125}{72}$.
The *diminished fourth* $\frac{4}{3} \times \frac{24}{25} = \frac{32}{25}$ and its inversion, the *augmented fifth* $\frac{3}{2} \times \frac{25}{24} = \frac{25}{16}$.
The *augmented fourth* $\frac{4}{3} \times \frac{25}{24} = \frac{25}{18}$ and its inversion, the *diminished fifth*, which is
   also called the *false fifth* $\frac{3}{2} \times \frac{24}{25} = \frac{36}{25}$.
An *augmented third* $\frac{5}{4} \times \frac{25}{24} = \frac{125}{96}$ and its inversion, the *diminished sixth* $\frac{8}{5} \times \frac{24}{25} = \frac{192}{125}$
are not in use.

14. Diatonic, Chromatic, and Enharmonic Progressions

(*Note: Names of tones in different languages.*) The scale mentioned in Par. 11, as also every other scale composed of major tones and semi-tones, is known as a diatonic scale, and the progression from one of these sounds to another contained in the same scale is known as a *diatonic progression*. Sometimes, the major semi-tone $\frac{16}{15}$ is the smallest degree on such a scale, the diatonic semi-tone. If one augments one of the sounds of the scale *do, re, mi, fa, sol, la, si* by the minor semi-tone $\frac{25}{24}$, this is expressed by the sign #, which is called a *sharp*; but if we lower the sound by the same interval $\frac{25}{24}$, it is expressed by the sign ♭, which is known as a *flat*. The sign of restoration is the *natural* ♮. A progression from a raised or lowered sound to the natural tone of the same denomination, or of the natural sound to the raised or lowered one, for example, from *do* to *do*$^{\#}$ or from *mi* to *mi*$^{\flat}$, is called a *chromatic progression*. Sometimes the minor semi-tone $\frac{25}{24}$, by which the progression of a raised sound to its neighboring lowered sound is made, is called the *chromatic semi-tone*. For example, from *do*$^{\#}$ to *re*$^{\flat}$ or from *re*$^{\flat}$ to *do*$^{\#}$ is called an *enharmonic progression*.

The origin of the names *do, re, mi, fa, sol, la, si* is too well known to be repeated here.[8] In different countries, the names of these sounds are different. In Italy, the syllables *do, re, mi, fa, sol, la* are used to express the degrees of any scale. In place of *si*, one then sings *mi* because it expresses the advance from the major semi-tone by *i:fa*; one then ordinarily changes the preceding syllable into a *re*. At the present time, some people add the syllable *si* because there are too many useless difficulties in wanting to express seven different objects by six signs. To express the same sounds, one makes use of the letters *C, D, E, F, G, A, B*, to which one adds the syllables that are suitable to the degrees of the ancient hexachord, in which this sound is found. Thus, for example, *do* is called *C sol fa do*, *re* is called D *la sol re*, etc. For elevation of the sound, a sharp is added, and for the lowering, a flat. In Germany, the sounds are called (beginning with *do*): *c, d, e, f, g, a, h* (which is pronounced *ha*). To express the sharp semi-tone, the termination *is* is added, saying *cis, dis, eis, fis, gis, ais, his* and to express the semi-tone flat, the termination *es* is added: *ces, des, es, fes, ges, as*; but to express *si^b*, one makes an exception, calling it *b*. One conforms more to the analogies of the other denominations if, as some have proposed, one wants to express the *si* by *b*, the *si$^\#$* by *bis*, and *si^b* by *bes*. One sees that the Italian denominations are more wordy, the German more precise.

The English and the Dutch call the sounds *c, d, e, f, g, a, b*. To express the semi-tone sharp, the English use the word *sharp* and the Dutch *kruis*, and to express the semi-tone flat, the English add *flat* and the Dutch *mol*.

15. Scales of Different Tones

All the sharp and flat intervals are necessary because, to avoid monotony, it is necessary that one be able to regard each sound as a fundamental sound, and to assign a fair scale to it. The series of sounds *do, re, mi, fa, sol, la, and si* does not contain all the degrees of these scales. If we regard the sound *sol*, for example as the fundamental, the sixth step to the seventh (*mi* to *fa*) will be only a semi-tone. It is

---

[8] It is popularly believed that Guido of Arezzo took the opening syllable letters from each line of a hymn to Saint John the Baptist to form the names of notes in the musical scale:

UTqueant laxis
REsonare fibris
MIra gestorum
FAmuli tuorum
SOLve polluti
LAbii reatum
Sancte **I**oanne—*RTB*

therefore necessary, in order that it be a tone, to use *fa#* in place of *fa*. In the same way, to use *re* as the fundamental sound, we must change *fa* into *fa#* and *do* into *do#*. For other fundamental sounds, we must flatten several sounds. For example, to have the fair scale of *fa*, it would be necessary to change *si* into *si♭* and to have the scale of this *si* it would also be necessary to change *mi* into *mi do♭♭*. Proceeding by fifths, it would always be necessary to sharpen one more sound, and in proceeding by fourths, or inverse fifths, it is also necessary to have a higher sound or a flattened one. One will therefore have the following diatonic scales:

do, re, mi, fa, sol, la, si, do	
sol, la, si, do, re, mi, fa#, sol	*fa, sol, la, si♭, do, re, mi, fa*
re, mi, fa#, sol, la, si, do#, re	*si♭, do, re, mi♭, fa, sol, la, si*
la, si, do#, re, mi, fa#, sol#, la	*mi♭, fa, sol, la♭, si♭, do, re, mi♭*
mi, fa#, sol#, la, si, do#, re#, mi	*la♭, si♭, do, re♭, mi♭, fa, sol, la♭*
si, do#, re#, mi, fa#, sol#, la#, si	*re♭, mi♭, fa, sol♭, la♭, si♭, do, re♭*
fa#, sol#, la#, si, do#, re#, mi#, fa#	*sol♭, la♭, si♭, do♭, re♭, mi♭, fa, sol♭*
do#, re#, mi#, fa#, sol#, la#, si#, do#	*do♭, re♭, mi♭, fa♭, sol♭, la♭, si♭, do♭*

These changes of the sounds in all the possible scales can be expressed by the arithmetic progression.

$$n^\# \ldots \ldots 3^\#, 2^\#, 1^\#, 0, 1^\flat, 2^\flat, 3^\flat \ldots \ldots n^\flat$$

If one wants to regard other sounds, for example, *sol#* or *fa♭* as the fundamentals, it is necessary to sharpen or flatten some sounds twice. When this becomes necessary, we express the double sharp by the sign **x** and the double flat by ♭♭.

The fundamental sound series *do, re, mi, fa, sol, la, si* (*c, d, e, f, g, a, b*) is the *range*, and the fundamental sound with the sounds that depend on it is the *mode*. If the fundamental sound has a major third, as in the series mentioned, it forms a *major mode*, if it has a *minor third*, it forms a *minor mode*.

16. Scale of the Minor Mode

To form the scale of the minor mode, we must give the perfect minor chords to the fundamental sound and to the sounds that approach it closer than the others, such as the fifth and the fourth. If we regard *la* as the fundamental sound, the perfect minor chord of this sound is *la, do, mi*; that of the fifth *mi, sol, si*; and that of the fourth *re, fa, la*. This will give the scale:

la, si, do, re, mi, fa, sol, la

But the ear requires that, in going up the scale, the step from the seventh to the eighth note be only a major semi-tone, which is called the *sensible note* (*subsemitonium modi*), because it determines every major or minor mode. It is therefore necessary, in going up the scale, to give to the fifth *mi* the major third *sol#*.

But, through this change, the step from the sixth note *fa* to the seventh *sol*$^{#}$ would be too large; it is therefore often necessary to use *fa*$^{#}$ instead of *fa*, and to regard the scale of the minor mode, in going up, as *la, si, do, re, mi, fa*$^{#}$, *sol*$^{#}$, *la*. This augmentation of the sixth and seventh notes is regarded as accidental, and they are pointed out every time they are used. In going down the scale, the scale remains unchanged.

Each scale of a minor mode contains the same notes as the major mode of its third minor; thus, for example, the scale of the minor mode of:

la is the same as that of the major mode:	*do*
mi...	*sol*
si...	*re*
fa$^{#}$...	*la*, etc.

17. Explanation of Several Words

When one mode contains more or less sharp or flat notes than the other, we say that they differ by so many *degrees*. A major and a minor that contain the same notes are *relative modes*. Sometimes we call the fundamental note the *tonic*, its fifth the *dominant*, its fourth the *subdominant*, and its third the *mediant*.

18. Progressions from One Chord to Another

The most natural progressions from one chord to another are that of the fifth or fourth, or to another which differs only by one degree. When one proceeds to more distant modes, one ordinarily makes by an enharmonic substitution of an augmented note to its neighboring diminished note, or from a diminished to its neighboring augmented, or one forces the ear to neglect the comma $\frac{128}{125}$ by which the major semi-tone $\frac{16}{15}$ differs from the minor $\frac{25}{14}$.

I will not develop any further the passages from one note to another or from one chord to another chord, since there are sufficient treatises on harmony that can provide instruction.

19. Relative Frequencies of Sounds Contained in an Octave

To provide a more exact idea of the size of each interval, I have furnished in the following Table the relative numbers of vibrations and the lengths of the corresponding chords, in fractions and in decimals, for each interval, reduced to the fundamental note *do*.

	Number of vibrations		Lengths of strings	
Unison, *do:do*	1	1	1	1
Minor semitone, *do:do$^{\#}$*	25/24	1.0416 2/3	24/25	0.96
Minor second or the major semi-tone, *do:re^{b}* .	16/15	1.066 2/3	15/16	0.9375
Major second, *do:re*	10/9	1.1111 1/9	9/10	0.9
(minor tone)				
or	9/8	1.125	8/9	0.888 8/9
(major tone)				
Diminished third, *do:mi^{bb}*	144/125	1.152	125/144	0.8680 5/9
(or, rather, *do$^{\#}$:mi^{b}*)				
Augmented second, *do:re$^{\#}$*	125/108	1.574 3/27	108/125	0.864
or	75/64	1.718 3/4	64/75	0.8533 1/3
Minor third, *do:mi^{b}*	6/5	1.2	5/6	0.8333 1/3
Major third, *do:mi*	5/4	1.25	4/5	0.8
Diminished fourth, *do:fa^{b}*	32/25	1.28	25/32	0.78125
Fourth, *do:fa*	4/3	1.3333 1/3	3/4	0.75
Augmented fourth, *do:fa$^{\#}$*	25/18	1.3888 8/9	18/25	0.72
Diminished fifth, *do:sol^{b}*	36/25	1.44	25/36	0.6944 4/9
Fifth, *do:sol*	3/2	1.5	2/3	0.6666 2/3
Augmented fifth, *do:sol$^{\#}$*	25/16	1.5625	16/25	0.64
Minor sixth, *do:la^{b}*	8/5	1.6	5/8	0.625
Major sixth, *do:la*	5/3	1.6666 2/3	3/5	0.6
Diminished seventh, *do:si^{bb}* (or *do$^{\#}$:si^{b}*)	128/75	1.7066 2/3	75/128	0.5959 3/8
or	216/125	1.728	125/216	0.5787 1/27
Augmented sixth, *do:la$^{\#}$*	125/72	1.7361 1/9	72/125	0.576
Minor seventh, *do:si^{b}*	16/9	1.7777 7/9	9/16	0.5625
or	9/5	1.8	8/9	0.5555 5/9
Major seventh, *do:si*	15/8	1.875	8/15	0.5333 1/3
Diminished octave, *do:do^{b}*	48/25	1.92	24/48	0.5208 1/3
Octave, *do:do*	2	2	1/2	0.5

Some people who are involved in practice have found fault with the theory in that it yields a minor semi-tone $\frac{25}{24}$, for example, *do* to *do$^{\#}$*, that is smaller than the major semi-tone $\frac{16}{15}$, *do* to *re*, although the minor sometimes has a better effect if one makes it slightly more acute; however, the theory is fair, and the reason why a minor semi-tone sometimes supports or requires slightly higher value is that ordinarily an augmented note rises to its more acute neighbor, and the ear likes to prepare and anticipate a little the tendency toward the following note.

20. Several Other Intervals Contained in the Natural Series of Numbers

The natural series of numbers still gives intervals that are not received in the ordinary system of sounds, and which, however, are produced by some musical instruments, such as the horn and the trumpet, where one must make use of such sounds for some others which they approach. The sound corresponding to the number 7, of which the effect is intermediate between the consonances and the dissonances, can be produced on these instruments, but they are not used. It would be useless to want to introduce them, because one would have to multiply too much the number of intervals which could scarcely be distinguished from those which already exist. It can, however, be presumed that the reason why the seventh chord (*do, mi, sol, si*) and the augmented sixth (*do, mi, sol, la$^{\#}$*) are also not disagreeable to the ear, which one could believe from their complex number, is due to the fact that the ear substitutes for these numbers the ratios 4:5:6:7, in which the interval $\frac{7}{4}$ differs from the seventh $\frac{16}{9}$ by the comma $\frac{64}{63}$, and from the augmented sixth $\frac{125}{72}$ by the still smaller comma $\frac{126}{125}$. In the same instruments, the sound corresponding to the number 11 is replaced by *fa*, but the interval $\frac{11}{8}$ is more acute by $\frac{33}{32}$ than the fourth $\frac{4}{3}$ or the true *fa*. Sometimes one makes it still more acute by employing more force and then we use it in place of *fa$^{\#}$*. The sound that corresponds to the number 13 is used for *la*, but the interval $\frac{13}{8}$ is graver by $\frac{40}{39}$ than the major sixth $\frac{5}{3}$. Sounds that surpass the number 16 are not used with horn or trumpet.

If one wishes to continue the natural series of numbers, even to infinity, one could never express certain intervals exactly, counting from the fundamental note; because there does not exist an integer to which some power of two were to be as 3 to 4 or as 5 to 6. However, the interval $\frac{9}{16}$ approaches the minor third $\frac{6}{5}$ closely, being no less than a comma $\frac{96}{95}$ away. Perhaps, when one sometimes uses the chord of the perfect minor chord in place of the major, or the major in place of the minor, the ear is less injured, because it substitutes for the minor third $\frac{6}{5}$ the interval $\frac{19}{16}$, thus hearing a variety of ratios like 16:19:24 and 16:20:24.

I would express these intervals in decimals in order to compare them with those numbers that were found in the previous paragraph.

Number of vibrations		Length of strings	
7/4	1.75	4/7	0.5714 2/7
11/8	1.375	8/11	0.7272 8/11
13/8	1.625	8/13	0.6153 11/13
19/16	1.1875	16/19	0.8157 17/19

Section 2: Altered Ratios or Temperament

21. The Necessity of Temperament

To judge the qualities and effects of sound, it is necessary to attribute to them the ratios mentioned above, but, for practical use, it is completely impossible always to make use of such ratios. If one wanted each progression of a sound to another to be fair, the ratio to the fundamental sound of the absolute pitch would not remain the same; but by assigning to each tone the fair value for the fundamental tone, they are not fair between them. A single example of a very simple succession of six tones: *sol, do, fa, re, sol, do* would suffice to make this clear. By making the procession of these sounds in the fair ratios and by expressing all these ratios by their least terms, we have:

$$\begin{array}{cccccc}
sol & do & fa & re & sol & do \\
(3{:}2) & (3{:}4) & (6{:}5) & (3{:}4) & (3{:}2) \\
243 : & 162 : & 216 : & 180 : & 250 : & 160
\end{array}$$

The *sol* appears once as 243 and another time as 240, and *do* appears once as 162 and another time as 160; one will therefore have lowered it by the comma $\frac{81}{80}$. If we wish to repeat this sequence of sounds, or if we want to execute any longer melody with exact ratios, we will be even more in error. If there are several voices that wish to continue their song with fair intervals, each will be incorrect in another way and there will no longer be any harmony. In wishing to carry out the preceding series of sounds in the exact ratios with the fundamental sound, we will have:

$$\begin{array}{ccccccc}
sol & do & fa & re & & sol & do \\
\frac{3}{2} & : 1 : & \frac{4}{3} & : \frac{9}{8} \left(\text{or} \frac{10}{9} \right) : & \frac{3}{2} & : 1
\end{array}$$

© Springer International Publishing Switzerland 2015
E.F.F. Chladni, R.T. Beyer, *Treatise on Acoustics*,
DOI 10.1007/978-3-319-20361-4_3

This process will also yield false ratios. If we take *do:re* as 9:10, the fourth *re:sol* will not be 3:4, but 20:27, too small by the comma $\frac{81}{80}$. But if we take *do:re* as 8:9, the minor third *fa:re* will not be 6:5, but 32:27, too small by the comma $\frac{81}{80}$. In trying any melody whatsoever in this manner, or if there are other sounds than these that belong to the perfect chord of the fundamental sound and to that of its fifth, we will have false results.

22. The 12 Real Notes

As it is not always possible to use exact intervals, it is at least necessary that each interval approach perfect exactitude as much as possible without deteriorating the others. The small alterations of the sounds that are necessary for this effect are called *temperament*. As each exact interval other than the octave is a little too large or a little too small for practical use, thus each minor semi-tone $\frac{25}{24}$ is a little too small and each major semi-tone $\frac{16}{15}$ is a little too large. It will therefore be necessary to execute the semi-tones, whatever their origin, as the main term between the minor $\frac{25}{24}$ and the major $\frac{16}{15}$; we will therefore have the 12 real notes:

do, do$^{\#}$/re$^{\flat}$, re, re$^{\#}$/mi$^{\flat}$, mi, fa, fa$^{\#}$/sol$^{\flat}$, sol, sol$^{\#}$/la$^{\flat}$, la, la$^{\#}$/si$^{\flat}$, si, do

which comprise the set of tones generally adopted.

> Some people are disposed to believe that temperament exists only for instruments with fixed sounds; but what we have said in Par. 21 will suffice to show that the bad results of too exact ratios are based on the nature of the ratios themselves. Every good singer, every good player of any musical instrument whatsoever tempers without knowing it.

23. The Results of Cycles of Fifths and Fourths

The most suitable intervals for determining the ratios of the 12 stops of this scale will be:

1. The fifths and fourths, because their ratio is the simplest after the octave, and because it is necessary that 12 fifths or fourths should yield an octave of the original note
2. The major thirds
3. The minor thirds, because it is necessary that three major thirds or four minor thirds should yield the octave of the fundamental sound

But the cycle of 12 fifths, *do, sol, re, la, mi, si, fa$^\#$, do$^\#$, sol$^\#$, re$^\#$* (or *mi^b*), *si^b, fa, do*, in the exact ratios of $\frac{3}{2}$ or of $\frac{3}{4}$, in order to have the sound in the same octave, gives, instead of the true octave 1:2, an interval of $2^{18}:3^{12}$ which surpasses the octave by the comma $\frac{531441}{524288}$, which is called the *Pythagorean comma*. The product of the 12 fourths, or $3^{12}:2^{20}$ gives an octave lowered by the same comma.

The *cycle* of the three major thirds, *do, mi, sol$^\#$, do*, gives the product $4^3:5^3$, which is smaller than the true octave by the comma $\frac{128}{125}$. The product of the *cycle* of four minor thirds, *do, mi^b, fa$^\#$, la, do*, or $5^6:6^4$ is larger than the true octave by the comma $\frac{648}{625}$.

24. Equal and Unequal Temperaments

It is necessary to start in some manner with a Pythagorean comma $\frac{531441}{524288}$ among the 12 fifths, the comma $\frac{128}{125}$ among the three major thirds, and the comma $\frac{648}{625}$ among the four minor thirds. All authors are in accord as to the necessity of this distribution, but they differ in their opinions as to the first manner of correcting the defects of the fifths and the other intervals. Some of us prefer an *equal temperament*, while others prefer an *unequal temperament*.

25. Preference for an Equal Temperament

It is an uncontested experience that if we hear one interval that differs very little from another that is expressible by simple numbers, we believe that we hear the simpler set, and that this illusion is the more perfect the smaller the difference. This illusion is very advantageous for us, since without it (Par. 21), there would be no music. For the effect, it is the same whether the interval one hears can be expressed by rational numbers or not. The end of each temperament being to distribute the difference in the least sensible manner, and the present state of music requiring us to be able to make use of each interval and each mode without hurting the ear; it follows that the equal temperament is the most conforming to nature. From the equal distribution of the difference over all the intervals, the octave excepted, the inexactness of each interval is too small to affect the ear in a disagreeable manner. All the homogeneous intervals are then of the same size, the 12 semi-tones that comprise the octave make a geometric progression, each fifth is lowered by the 12 parts of the Pythagorean comma, each major third is raised by the third part of the comma $\frac{128}{125}$, and each minor third is lowered by the fourth part of the comma $\frac{648}{625}$. No interval is harmed by another, because, if a tone is fairly tempered like a fifth, it has also the fair ratio as the major and minor thirds.

26. Calculation for Equal Temperament

The calculation for equal temperament or for the geometric progression of the numbers between 1 and 2 can be done in different ways. One of the simplest is the following:

We divide the octave *do:do* into two equal intervals, taking the geometric mean between 1 and 2, which yields the square root of 2, or 1.41411 for the tone $fa^\#$ or sol^b. We also divide it into three equal intervals, in order to have the major thirds *do : mi : sol^\# : do*. The two geometric means between the two numbers, of which I express the one by p and the other by q, are:

$$p : \sqrt[3]{p^2 q}: \quad \sqrt[3]{q^2 p} : q;$$ or p here being $=1$, and $q=2$, the cube root of 2 or 1.25992...gives the tone *mi*, and that of 4 or 1.58740...the tone $sol^\#$ or la^b.

These numbers suffice to find all the others and for that we have only had to take the square root of the product of the two numbers, between which we wish to find the new tone. The square root of the product of:

$$\begin{aligned}
do \ \text{and}\ fa^\# \ \text{will give}\ re^\# &= 1.18921 \\
fa^\# \ \text{and}\ do \ldots\ldots ..la &= 1.68179 \\
do \ \text{and}\ mi \ldots\ldots . .re &= 1.2246 \\
sol^\# \text{and}\ do. \ldots\ldots . si^b &= 1.78180 \\
do \ \text{and}\ re \ldots\ldots . . do^\# &= 1.05946 \\
mi \ \text{and}\ fa^\#. \ldots\ldots . fa &= 1.33484 \\
fa^\# \ \text{and}\ sol^\#. \ldots\ldots sol &= 1.49831 \\
si^b \ \text{and}\ do \ldots\ldots . .si &= 1.88775
\end{aligned}$$

One therefore has the following sequence of sounds, to which I have added the lengths of the corresponding strings:

Number of vibrations			Length of strings		
do	=	1.00000	*do*	=	1.00000
$do^\#$ or re^b	=	1.05946	$do^\#$ or re^b	=	0.94387
re	=	1.12246	*re*	=	0.89090
$re^\#$ or mi^b	=	1.18921	$re^\#$ or mi^b	=	0.84090
mi	=	1.25992	*mi*	=	0.79370
fa	=	1.33484	*fa*	=	0.74915
$fa^\#$ or sol^b	=	1.41421	$fa^\#$ or sol^b	=	0.70710
sol	=	1.49831	*sol*	=	0.66742
$sol^\#$ or la^b	=	1.58740	$sol^\#$ or la^b	=	0.62996
la	=	1.68179	*la*	=	0.59461
$la^\#$ or si^b	=	1.78180	$la^\#$ or si^b	=	0.56123
si	=	1.88875	*si*	=	0.52973
do	=	2.00000	*do*	=	0.50000

Another method, which is essentially the same, and which gives the same results, consists in multiplying 12 times in succession with the square root of 2, which can be done better with algorithms than with the numbers themselves. We express this geometric progression by:

$$do : do^{\#} : re : re^{\#} : mi : fa : fa^{\#} : sol : sol^{\#} : la : si^{b} : si^{\#} : do$$

$$1 : 2^{\frac{1}{12}} : 2^{\frac{2}{12}} : 2^{\frac{3}{12}} : 2^{\frac{4}{12}} : 2^{\frac{5}{12}} : 2^{\frac{6}{12}} : 2^{\frac{7}{12}} : 2^{8/12} : 2^{\frac{9}{12}} : 2^{\frac{10}{12}} : 2^{\frac{11}{12}} : 2$$

Or by:

$$1 : \sqrt[12]{2} : \sqrt[12]{2^2} : \sqrt[12]{2^3} : \sqrt[12]{2^4} : \sqrt[12]{2^5} : \sqrt[12]{2^6} : \sqrt[12]{2^7} : \sqrt[12]{2^8} : \sqrt[12]{2^9} : \sqrt[12]{2^{10}} : \sqrt[12]{2^{11}} : 2$$

Thus each interval of our system, except the octave, cannot be rigorously expressed in terms of irrational numbers which always represent other simpler ones, from which they differ only in an almost imperceptible manner to our sense. Thus, the fifth $\sqrt[12]{2^7}$ differs from the true $\frac{3}{2}$ only by the comma $\frac{149831}{150000}$ and the major third $\sqrt[12]{2^4}$ differs from the true $\frac{5}{4}$ only by slightly less than the comma $\frac{125}{126}$. If one were to assign more exactness to the interval, the other intervals would worsen.

27. Practical Application

In tuning instruments, it would suffice to lower each fifth and raise each major third almost imperceptibly. We will therefore always have a better temperament than if we execute it by a design of several intervals that are more exact than the others, or if we wish to make several intervals in the opposite sense. The ear can still support some fifths lowered by slightly more than a 12th part of the Pythagorean comma, but $\frac{2}{12}$ or $\frac{2\frac{1}{2}}{12}$ of the same comma will be the limit of the supportable fifths.

28. Rules for Judging Unequal Temperament

Just as there is only one truth and an infinity of errors, just so there is only one equal temperament, but as many unequal temperaments as one can wish. Here are principles for judging their relative value or, if one wishes to put it that way, their defects.

1. The more exact the fifths, the poorer is the temperament, because then this small number of fifths, among which the Pythagorean comma is distributed, become less supportable.
2. The same results follow if the Pythagorean comma is more unequally distributed.
3. The worst temperaments are those where there are raised fifths, because then several other fifths support the excess of the raised fifths, besides the Pythagorean comma.

The temperament of Kirnberger[1] is one of the worst because it contains nine
exact fifths and the Pythagorean comma is distributed unequally over three
fifths. It must be noted here that, because of the authority of Kirnberger,
otherwise justly celebrated as a harmonist, this has resulted in several authors
adopting false principles.

L. Euler (*Tentamen novae theoriae musicae*; *Nov. Comment. Acad Petrop.*
vol. XVIII and *Mém. de l'Acad. de Berlin*, 1764) expresses the series of
12 tones contained in the octave $2^m 3^3 5^2$, multiplying all the divisors of 3352
sufficiently often by 2 in order to return to the same octave. One will have,
therefore, the series of sounds $do = 384$, $do^\# = 400$, $re = 432$, $re^\# = 450$,
$mi = 480$, $fa = 512$, $fa^\# = 540$, $sol = 576$, $sol^\# = 600$, $la = 640$, $si^b = 675$,
$si = 720$, $do = 768$. This sequence of tones approaches the true values more
closely than any other expressible by irrational numbers, but it is not appli-
cable in practice because the fifth si^b:fa is too sharp by the comma $\frac{2048}{2025}$ or $\frac{10}{12}$ of
the Pythagorean comma. The sum of the differences of the fifths will then be
$\frac{22}{12}$ of this comma which is distributed over the fifths re:la and $fa^\#$:$do^\#$, lowered
by $\frac{81}{80}$; four major thirds $do^\#$:fa, $re^\#$:sol, $sol^\#$:do, and si^b:re are too sharp by the
comma $\frac{128}{125}$, etc.

It would be superfluous to examine anything that is as useless and
disagreeable as unequal temperaments, proposed by several authors,
where each one pretends that this method of tempering is preferable to all
the others.

The best work on temperament that I know, and from which I have
borrowed several ideas, is that of Marpurg's *Versuch ber die musikalische
Temperatur* (also called *Treatise on Musical Temperament*), Breslau, 1776.

Appendix to Part I

29. Signs for the Tones Contained in Different Octaves

It is necessary for me to have *signs for the tones contained in different octaves*
because, in Part II, I will give series and tables for the tones that the same elastic
body can produce in its different modes of vibration. I will therefore regard the
lowest *do* of the clavier or violoncello as the basis. I will express each tone of
this first lower octave by adding the number 1, for example, *do 1*, $do^\#$ *1*, *re 1*.

---

[1] Johann Philipp Kirnberger (1721–1783), composer, musical theorist, and pupil of Johann
Sebastian Bach, created a tuning system based on equal temperament.—*MAB*

The tones of the second lower octave will be expressed by adding the number 2. For the tones of the following octave, which is the first above, we will add the number 3, for the second above, the number 4, and so on. When it is necessary to mention a tone that is lower than the first *do* of the clavier or violoncello, I will express it by a minus sign placed as a subscript. By adding the plus sign I will express that the sound is slightly more acute than the tone mentioned.

As far as I know, no names or signs exist that are generally accepted for the tones contained in different octaves, except in German, where one expresses tones that are graver than the first *do* of the violoncello by a line placed under the name; the tones contained in the first lower octave, beginning with the *do*, by the initial letters; those of the following octave by the ordinary letters, those of the first octave above by a line written above, those of the second octave above by two lines, etc.

Part II
Characteristic Vibrations of Sounding Bodies

Section 1: General Remarks

30. Different Types of Sounding Bodies

The preceding part concerned the frequencies of vibrations in general; but here it will be a question of the nature of the vibratory motions with regard to changes in shape and corresponding frequencies in each kind of motion of any sounding body whatsoever.

Elasticity being the moving force for sonorous vibrations, a sounding body can be *elastic*, either under *tension*, or under *compression*, or under due to *shear* (stress).

Flexible bodies that become elastic under *tension* can be either *thread-like*, when changes in the shape can be expressed by curved lines as in *string*s, or *membrane-like*, where the changes in shape can be expressed, not by curved lines, but by curved surfaces, as in the membranes of drums (timbales) and other *taut membranes*.

Examples of sounding bodies that are elastic under *compression* are the air and the gas in wind instruments.

Bodies that are elastic by virtue of their internal *stiffness* are either thread-like or membranous. The thread-like bodies can be either straight-line, like rods or sheets, or curved, like rings, forks, etc. Rigid membranous bodies are also either rectilinear, as in plates of any shape whatsoever, or curved, as in vessels and bells.

In this manner of regarding sounding bodies, there are none that cannot be reduced to one of these types. The vibrations of most of these bodies were wholly unknown; but I have made use of new means of making them sensible to the eye and ear.

I will add some remarks on the coexistence of several motions in the same sounding body.

© Springer International Publishing Switzerland 2015
E.F.F. Chladni, R.T. Beyer, *Treatise on Acoustics*,
DOI 10.1007/978-3-319-20361-4_4

31. On Noise and on the Different Timbres of Sound

In distinct sound, the vibrations of a sounding body or of its parts are carried out at the same time, and all the vibrations are of equal duration; but one cannot say the same thing about *noise*, the nature of which is still not known. Sound, when the mode of vibration, the frequency, and the force are the same, sometimes has a very different character, which we call *timbre*. This seems to depend on the different stiffness or tenacity of the bodies, and on the quality of the material that serves to set them into motion. We do not know the real causes of these different effects, and there are still no means of submitting them to calculation or experiment. This difference of timbre seems to be caused by a little noise mixed with the perceptible sound. For example, in song, one hears, in addition to the vibrations of the air, the rubbing of the air on the organs of the voice. In the violin, in addition to the vibrations of the strings, we hear the rubbing of the bow on the strings. Perhaps the different types of noise and timbre consist of unequal motions of the smallest parts of the body, as those by means of which Lahire, Carr, and Musschenbroek wished to explain the nature of sound. But, instead of continuing to make conjectures about the nature of the noise, and of the different timbre of these sounds, we will explain the nature of perceptible sound.

32. General Laws of Sound Vibrations

Every sounding body can undergo vibrations of very different types, the character of each of which has a certain ratio to the frequency of the others, depending on the size of the vibrating parts.

When the sounding body is divided into a number of vibrating parts, these parts (of which the oscillations are called the loops or antinodes of the vibrations and are separate by fixed limits which are called the nodes) always make their motions alternately in the opposing sense, in such a way that one is just above the rest position, while the next is just below it.

The isochronism of the vibrations of all the parts produced by their relative equilibrium being an indispensable condition for the sound, it must, like the division of the sounding body into vibrating parts, be always as regular as the circumstances permit. The size of a part situated at the free end is about half that of the part that is located between the nodes.

In order to produce a certain sound, we can hold or touch one or more of the nodes and rub or strike one vibrating part in the same direction as the vibrations are made.

Several or all of the ways of vibrating can coexist in the same sounding body. The body motions can also coexist with other forms of motion.

This paragraph is distinguished from the others because it contains a précis of all the laws of sound vibrations. All the rest of acoustics is only showing the ways in which these same laws are differently modified in different elastic bodies.

33. The Vibrations Must Be Very Small

For the theory, it is assumed that the vibrations of a sounding body, like the oscillations of a pendulum, are *infinitesimally small*, but, when they are really only *very small*, the difference in the calculations of the vibrations that results from that fact is not considerable. If it is a question of a taut string which makes oscillations of one degree, the duration of one vibration is greater by $\frac{1}{30000}$, if the oscillations were infinitely small. For an oscillation of two degrees, the difference is approximately $\frac{1}{12500}$, etc.

I have represented much greater oscillations in the figures in order to distinguish them more easily.

34. Different Directions of Vibration

The *direction* of the vibratory motion can be *transverse*, *longitudinal*, or *torsional*.

In *transverse vibrations*, the sounding body or the parts of this body make their motions alternately toward one side or the other in such a way that the lines traveled by each point of the body make a right angle with the axis. Figures 1–4, 20–27, and 38–40 serve to give an idea of similar motions.

Longitudinal vibrations consist of contractions and expansions of the sounding body or of its parts in the direction of the axis, or along the length, as in Figs. 14–19 and 28–36. Bodies susceptible to such motions include:

1. The air contained in a wind instrument
2. Strings or rods of sufficient length

The laws of these two types of vibrations are very different.

Torsional vibrations, to which some rods or plates are susceptible, consist of rotations which occur alternately in opposed senses. In cylindrical or prismatic rods, the sound of these vibrations is always graver by a fifth than the longitudinal sound of the same body, driven in the same manner.[1]

---

[1] We now know that the ratio of torsional wave velocity to compressional wave velocity depends on the shear modulus, the Young's modulus, and the Poisson ratio for the material. It is not clear what Chladni intended by "driven in the same manner."—*TDR*

35. Intensity of Sound

The intensity of sound depends on the magnitude of the oscillations, on the size of the sounding body, and on the frequency of its vibrations.

Section 2: Vibrations of Strings

A. Transverse Vibrations

36. Modes of Vibration

A *string* can vibrate as a whole or divided into any number of equal segments, separated from one another by nodes of the vibrations. The only difference between these vibrations is that the unity which serves as a measure changes, because, when the string is divided into aliquot parts, each half, each third, etc. makes its motions as if it were a string by itself. The gravest sound is that made when the entire string vibrates and alternately forms the loops represented in Fig. 1 by *ACB* and *ADB*. When it is divided into two parts, one half is on one side of the rest position, while the other is on the opposite side, and the loops are as in Fig. 2, *ADCEB* and *AFCGB*; the sound is more acute by one octave than the fundamental. If the string is separated into three segments, the loops are alternatively placed, as marked in Fig. 5, in two different ways; the sound is more acute by a fifth than that of the second harmonic. If the string is divided into four segments (Fig. 4), the pitch of the sound is increased by a fourth. In general, all of the possible sounds are as the numbers of segments (or as the reciprocals of their lengths); the series will therefore be as the numbers 1, 2, 3, 4, etc. When the gravest is *do*, the series of possible sounds will be:

No. of segments	1	2	3	4	5	6	7	8	
Sounds	*do 1*	*do 2*	*sol 2*	*do 3*	*mi 3*	*sol 3*	*si^b 3−*	*do 4*	
	9	10	11	12	13	14	15	16	etc.
	re 4	*mi 4*	*fa 4+*	*sol 4*	*la 4−*	*si 4−*	*si^b 4−*	*do 5*	

In a string of unequal thickness, the vibrations are ordinarily very irregular, except for several special cases; for example, if the lengths of the segments are in inverse order to their diameters.

© Springer International Publishing Switzerland 2015
E.F.F. Chladni, R.T. Beyer, *Treatise on Acoustics*,
DOI 10.1007/978-3-319-20361-4_5

37. On the Manner of Producing These Vibrations and Making Them Visible

To produce sound where the string is divided into aliquot segments, one must place a finger very lightly at a point where there is a node of vibration, and apply a violin bow approximately in the middle of the vibrating part. It is not necessary to press the node of vibration very hard, in order to prevent the transmission of the motion of one segment to another; the pressure of the bow should also be much smaller than that for the fundamental sound. The mode of division can be made visible by placing small bits of paper on different points of the string; those which are on the vibrating part will be struck by the vibrations and fall off; but those which are placed on the node of vibrations will remain stationary.

We owe this experiment *to Sauveur* (*Hist. et Mém. de l'Acad. de Paris*, 1701). Wallis (in *Algebra*, vol. 2, p. 466) mentioned sounds of aliquot segments as a discovery made by Noble and Pigot at Oxford, and communicated to him in 1676 by Narcissus Marsh.

The sounds of aliquot parts of a string on the violoncello and on the violin are sometimes used. These are known as *fluted sounds* or *harmonic sounds*. Use is also made of it in the case of an instrument with a single string, which is called a *marine trumpet*. The sounds of the *Aeolian harp* consist of similar vibrations, produced by an air current that acts on the strings in different ways. Ossian and the commentator on Homer, Eustathius, had already mentioned the sound of strings produced by the wind. A. Bale, in the house of a Captain Haas, had very long and very strong strings, exposed to the air, that yielded different sounds, especially during changes in intervals. In the *Annali di Chimica e storia natural*, (Pavia), vol. 18, 1800, similar observations were made by Gaetano Berrettari.

38. Coexistence of Several Vibrations

Several or all kinds of vibrations to which a string is susceptible can exist at the same time. In order to have an idea of the loops of the string, it is not necessary to apply a curve to a straight line, but to the curve which already exists at each moment by other vibrations. Figs. 5–8 represent several examples of similar curves. Sect. 9 of this Part will contain more instruction on this subject.

39. On the Curve Formed by a String in Its Transverse Vibrations

The opinions of geometricians differ on the nature of the curves into which a string can convert itself in transverse vibrations. Taylor, D. Bernoulli, and Giordano Riccati have found that these curves have the shape of a very elongated trochoid and that L signifies the length of the string, the ratio of the periphery of the circle to its diameter. And if one expresses the greater order at the middle of a segment of vibrations for the first kind of vibration by A, for the second by B, for the third by C, and so on, any abscissa by x and the number ordered to this abscissa by y for the fundamental type of vibrations, y is estimated to be $= A \sin \dfrac{\pi x}{L}$, for the second $y = B \sin \dfrac{2\pi x}{L}$, for the third $y = C \sin \dfrac{3\pi x}{L}$, etc. But Euler presumed that the wave is arbitrary and that it depends on the first impressions that are made on the string, in the manner that there will not always be continuity of the different segments of the curve, but that each vibrating part takes on the same curve as the other alternatively in the opposite sense. Lagrange has proposed the same opinion as Euler. D'Alembert also attributes still other curves to the strings, such as the trochoidal curves of Taylor, but he did not agree that the string could take on curves which do not conform to any law of continuity.

40. Laws of These Vibrations

If L expresses the length of the string, G the weight, P the tension (which can be expressed by a suspended weight), n the number of segments into which the string is divided and S the relative number of vibrations, or the sound of the string, S will be $= n\sqrt{\dfrac{P}{LG}}$. In strings that are made of the same material, if D expresses the diameter or thickness, G is $= D^2 L$ and $S = n\sqrt{\dfrac{P}{L^2 D^2}}$, or $\dfrac{\sqrt{P}}{LD}$. Consequently, homogeneous sounds (where n is the same) of strings of the same material will be:

1. When the thickness and the tension are the same, the sounds will be as the *reciprocal lengths* of the strings; that is why we can use a monochord for demonstration of the ratios of the sounds.
2. When the length and the tension are the same, m, the sounds will be as the reciprocal of the *diameters* (or *the square root of the weight*); in such a way that if, for example, the thickness of one string is to that of another as 1–2, the sound of the thicker string will be lower by an octave.

3. When the thickness and the length are the same, the sounds will be as *the square roots of the tension*. If one wishes, for example, for the sound of a string to differ from that of another by an octave, it is necessary that the tensions be as 1–4.

Differences in material have no affect on the determination of sound; a string of catgut and a string of any metal whatever will give the same sound if the length, the weight, and the tension are all the same. The duration of each vibration being reciprocal to the number of vibrations, it will be $\frac{1}{n}\sqrt{\frac{LG}{P}}$. The absolute number of vibrations that the string makes in a second of time can be found by comparing it to a seconds pendulum, where the duration of a vibration is expressed by π (or the ratio of the circumference of the circle to the diameter) multiplied by the square root of the length. The length of a seconds pendulum being f; a second, or the duration of an oscillation of the pendulum, will be at t, or to the duration of a vibration of the string as $\pi\sqrt{f}$ is to $\frac{1}{n}\sqrt{\frac{LG}{P}}$; t will be equal to $\frac{1}{\pi n}\sqrt{\frac{LG}{f\,P}}$, and the number of vibrations that are made in a second $\pi n\sqrt{\frac{fP}{LG}}$.

41. Authors Consulted

To learn more about transverse vibrations of strings, research on the vibrations of strings can be found in the work of Brook Taylor (*Methodius incrementorum directa et inversa*, London, 1715) where one finds the best analytic research on the vibrations of strings; Johann Bernoulli (*de chordis vibrantibus, Comment. Acad. Petrop.*, vol. 3); Leonhard Euler (*Mém. de l'Acad. Berlin*, 1748, 1753, and 1765; *Nov. Comment. Acad. Petrop.*, vol. 9, 17, and 19; *Acta Ac. Petrop.*, 1779, p. 2; 1780, p. 2 and 1781, p. 1; *Mélanges de Philosophie et de Mathématiques de la Société de Turin*, vol. 3.); Daniel Bernoulli (*Mém. de l'Acad. Berlin*, 1753 and 1765; *Nov. Comment. Acad. Petrop.*, vol. 16.); J. Lagrange (*Mélanges de Philosophie et de Mathématiques de la Société de Turin*, vol. 1, 2, and 3); D'Alembert (*Mém. de l'Acad. Berlin*, 1747, 1750, and 1763; *Opuscul.* vols. 1 and 4); Giordano Riccati (*delle corde ovvero fibre elastiche*, Bologna, 1767); Matthew Young (*Enquiry into the principal Phenomena of sounds and musical strings*, Dublin, 1784); Zanotti (*de vi elastica*, in *Comment, Bonon.*, vol. 4).

42. A Particular Case: Where the Tone from a String Divided into Two Parts Is Lower than That of the Entire String

I will now add a unique phenomenon, in which a string, divided into two parts, produces a lower tone than that corresponding to the ordinary vibrations of the entire string. Hellwag, physician to the reigning Duke of Oldenburg at Eutin, having observed it, was kind enough to communicate it to me. If we place a support under the string, so that it is not fixed but lightly touched, and if we pinch the string in order that it strike vertically in the support, there will be some cases in which one will hear the striking as a perceptible sound lower than its fundamental, but very raucous and disagreeable because of the deformity of the vibrations. This sound can be called the "snoring sound" of the string.[1] If we apply the support to the middle of the string, the snoring sound is graver by a fifth than the ordinary sounds of the entire string. When the string, Fig. 9, is pulled from its rest position pnq toward m and released, it strikes the support n after one half of a vibration; the two halves continue their motions forming the curves pkn and nfq; then they return and, as soon as they reach the axis pnq the entire string makes a half **vibration toward pmq and another toward pnq**, and so on. One therefore hears the shocks on the support in the sum of the following time intervals:

1. The half vibration of each half pn and nq toward pkn and nfq, one quarter of an ordinary vibration
2. The return of each half to the axis pnq, one quarter of a vibration
3. The motion of the entire string toward pnq, half a vibration
4. The return of the entire string to the axis, where it strikes the support, one half a vibration

The time interval, therefore, between two strikings of the support, $\frac{1}{4}+\frac{1}{4}+\frac{1}{2}+\frac{1}{2}=\frac{3}{2}$; it must be that the snoring sound is a fifth lower than the ordinary sound, confirming to the experiment. But, because of the motions of each half, there is always a mixture of the higher sound that belongs to these halves, and finally, when the shocks cease, the more acute sound continues for a bit. There are only two cases which find this sound perceptible, but much less distinct. If the string is divided in the same way into two parts, which are $\frac{2}{5}$ and $\frac{3}{5}$, the snoring sound is a semi-tone more acute than in the previous case. It seems to me that the ratio to ordinary sound is $\frac{18}{25}$ to 1. If the support divides the string into two parts that are $\frac{1}{3}$ and $\frac{2}{3}$, the snoring sound is lower by a ninth than the ordinary sound. The ratio is

---

[1] The "snoring sound" refers to the raucous and disagreeable sound caused by the string striking its support.—*GB*

therefore $\frac{2}{9}$ to 1. The effect was almost the same if the end where one plucks the string was different, or if the support was not placed exactly in the place mentioned.

B. Longitudinal Vibrations of Strings

43. Different Types of Longitudinal Vibrations

A longitudinal vibration consists of the contraction and expansion of a string, or its aliquot parts, moving alternately between fixed points or a vibrational node. In the simplest longitudinal motion, the entire string moves alternately toward one fixed point and then toward the other (Figs. 34a, b). The second kind of longitudinal motion is that in which the string is divided into two equal parts, which alternate between the vibrational node in the middle and the fixed points at the ends (Figs. 35a, b). In the third type of longitudinal vibrations, the motions of the parts are alternately as Figs. 36a, b, etc. The sounds together have the ratios as those of transverse vibrations, being as the numbers 1, 2, 3, 4, etc.; but there is no fixed relationship for the absolute pitch of the sound between these two types of motion because the laws are very different.

44. Manner of Producing Them

To produce these sounds, we must rub a vibrating part of the string longitudinally with a violin bow, which is held at a very acute angle, or with a finger, or with another flexible body to which one has applied rosin powder. For division of the string into aliquot parts, it is necessary to touch a node of the vibrations lightly at the same time.

45. Laws of These Vibrations

The laws of longitudinal vibrations differ altogether from those of the transverse vibrations. The only resemblance is that the sounds are in the inverse ratio to their lengths; but in longitudinal vibrations, the sound does not depend on the thickness of the string or on the tension, but only on the length and the type of material of which it is made. For example, a brass string gives a more acute sound by about a sixth than that of a catgut string. And the sound of a steel string surpasses the one of the brass string by about a fifth. To do the experiments, it is necessary to use strings of a considerable length, since these sounds are very acute. I even used strings that were 48 ft long. Sect. 5 of this Part will contain more information on longitudinal vibrations.

Section 3: Vibrations of a Stretched Membrane

46. Explanations

A *rectangular taut membrane,* only stretched lengthwise, will be susceptible to the same vibrations and the same sounds which vibrate crosswise. The nodes of vibration will then be transverse immobile lines. But such a membrane, as a membrane stretched in more directions, will also be able to vibrate in infinite ways. The curvature cannot be expressed by lines, but by curved surfaces. For these, the expressions and the means to calculate them are once again lacking. The nodes of vibration will form nodal lines in the different directions. More information on the vibrations of surfaces is located in Sect. 7.

If rectangular membranes vibrate as a string, it will be necessary (if the material is the same) to change the expression (Par. 40) $\frac{\sqrt{P}}{LD}$, if B signifies the width, to $\frac{n}{L}\sqrt{\frac{P}{DB}}$.

47. Modes of Vibration

According to the research of Giordano Riccati *(Saggi scientifici e letterari dell' Academia di Padova,* vol. 1, *1786, p.* 414ff) on the vibrations of a membrane of a kettledrum stretched equally in all directions, some vibrations correspond to those of strings, giving the same ratios of the sounds. A kettledrum of which the gravest sound is *si*^{b} *1,* also gives the sound *la 2,* more acute by almost an octave. And the sound *mi 3,* even more acute by a fifth. If L expresses the diameter, M the mass of the membrane, P the tension, π the ratio of the circumference to the diameter, $\mathfrak{f}$ the number of seconds, and n the number of the vibrations which accompany each type of motion, the number of vibrations in one second of time will be, according to Riccati, $= \frac{3}{4}\pi n\sqrt{\frac{\mathfrak{f}\,P}{LM}}$.

© Springer International Publishing Switzerland 2015
E.F.F. Chladni, R.T. Beyer, *Treatise on Acoustics,*
DOI 10.1007/978-3-319-20361-4_6

Riccati's assumption, that in the vibrations of such a kettledrum membrane, every diameter can bend itself to the curves of a vibrating string, is true for the fundamental sound and for all the sections of a string divided into an odd number of parts; but it is impossible for the sections of a string divided into an even number of parts. To prove this, I will express every elevated section in the ordinary position by + and every section lowered under this position by −, as one does for any type of opposing quantities. When such a membrane gives the exclusive fundamental sound, every diameter takes the curvature of a string in simpler vibrations (Fig. 1). But if one wants to suppose that every diameter of the membrane can move as a string divided into two parts (Fig. 2), it is necessary (Fig. 10) that at the same time: am is + and bm −, cm is + and dm −, em + and fm −, gm + and hm −, bm + and am −, dm + and cm −, etc. It is necessary, therefore, that every semi-diameter is at the same time above and below the ordinary position.

Consequently, such a supposed mode of vibration, where each diameter gives the same motion as a string in Fig. 2, will not exist. But it will be represented by another, where the membrane (Fig. 10) will be divided by a nodal line ef into two half-round parts, eaf and fbe, of which one is +, while the other is −; and where only the diameter amb vibrates exactly as a string in Fig. 2, and every other diameter in a slightly different manner; and where the diameter ef doesn't vibrate; these types of vibrations will not be able to be described with a curved line.

The third mode of vibration of a string (Fig. 5) will apply to every diameter at the same time; the nodal lines will then form a concentric circle (Fig. 11).

The fourth curvature of a string (Fig. 4) will not apply to every diameter at the same time, for the same reason that excludes the second; but a single diameter (Figs. 12a, b) will take this curvature, and the nodal lines will be a circular line and a diametral. The curve of a string divided into five parts will apply to every diameter, and the nodal lines will form two concentric circles (Fig. 13, etc.). Besides these types of vibrations analogous to the vibrations of a string, the membrane will be able to be divided in many other ways, where there is more of a nodal line in diametral directions, etc.

Euler has published, in Vol X of the *Nov. Comment. Acad. Petrop.*, some research on the vibrations of a rectangular membrane. In the fourth volume of *Mémoires de Mathématiques et Physiques de l'Institut de France*, Mr. Biot calculated the possibilities for the division of a taut rectangular membrane into aliquot parts. There are not yet experiments on this subject. It will be necessary to find a new means to do them, because in similar membranes, the edge is not free, and one cannot apply a violin bow.

Section 4: Vibrations of the Air in a Wind Instrument

48. The Air Here Is the Sounding Body

Here, it will be the question of the vibrations in the air, when it is itself the sounding body. That which is transmitted in the air by another sounding body will be reserved for Sect. 1 of Part III. It will be regarded as the continuation of this one, the laws being the same.

49. A Simple Blow Is Enough to Produce Vibrations

Every strong simple blow, for example, the crack of a whip, or an explosion, produces vibrations in the air. But they are ordinarily too irregular and with too few isochrones to give a distinguishable sound.

50. Sound Produced by a Current of Air Through a Slit

The vibrations of the air produced by the passage of a current of air through an opening or narrow slit are much more distinguishable. The frequency of these vibrations is dependent on:

1. The speed of the air current: If the opening remains the same and the speed of the air current increases, the sound is more acute.
2. The size of the opening: If the opening is smaller and the speed of the air current remains the same, the sound is more acute.

If both the speed of the air current and size of the opening increase or decrease, the sound will remain the same, but the intensity will be different. The whistling

© Springer International Publishing Switzerland 2015
E.F.F. Chladni, R.T. Beyer, *Treatise on Acoustics*,
DOI 10.1007/978-3-319-20361-4_7

from the compression of the levers, and the sound that the wind produces sometimes when it is passing through a narrow slit, will serve as examples. The sounds that one can produce while blowing, for example, an oboe reed, follow the same laws.

51. Reinforcement of the Sound of the Air by the Resonance of a Membrane

If a current of air passes through an opening or a narrow slit, forcing a membranous body to vibrate, the sound is greatly reinforced, but typically it becomes more piercing or more breathy. This is what takes place, for example, if one stretches a small piece of paper, a blade of grass, or a reed between the thumbs of both hands, and blows air over both sides of the object. Such a reinforcement by a vibrating membrane for a pipe with reeds also applies to pipe organs.

52. Voices of Humans and Animals

The *voices of humans and animals* are formed in the same way. Two roughly half-round membranes are found in the larynx, which together form a circular surface. The outer edges of the membranes, which are called the *ligaments of the glottis*, are attached to the walls of the larynx, and their straight edges are able to fold along the diameter of the circle to form a lens-like split that is called the *glottis*. If this opening is too large, air passes without producing a sound, as with ordinary respiration. But if it is drawn closer or narrowed, the air that is passed out of the lungs through the larynx rubs against these two membranes and produces rapid vibrations that are transmitted to the air going out. To this current of vibrating air, that is called the *voice*, other organs of the mouth present different obstacles and very differently shaped openings, each of which varies and articulates the voice in a different way.

The more the glottis is narrowed by the tension of the ligaments, the more acute the pitch. All of the possible varieties are produced by changes to the opening, where the most extreme difference is only a tenth of an inch.

According to Dodart (*Mém. de l'Acad. de Paris*, 1700, 1706, and 1707), the different sounds depend on the widening and narrowing of the glottis; but Ferrein (*Mém. de l'Acad. de Paris*, 1741 and 1743), claims that it depends on the differences in the tension of the ligaments. But the two assertions are not contrary, because when the opening is narrowed, the ligaments are stretched.

(continued)

The best information on the vocal organs of birds (where they are more complex), mammals, and reptiles appears in the *Leçons d'Anatomie comparée* by Cuvier, vol. iv, lesson xxvii. One may also read Haller, *de partium corporis humani fabrica et fonctionibus*, Book IX; Vicq d'Azyr, *sur la voix*, in the *Mem. de l'Acad. de Paris*, 1779; Ballanti, Urtini, and Galvani *observationes de quorundam animalium organo vocis* in *Comment. Bonon.*, vol. VI, p. 50ff.

Von Kempelen of Vienna has published much interesting research in his book, *Über den Mechanismus der menschlichen Sprache* (*On the Mechanism of Human Language*), in Vienna, 1791, where an exact description of the speaking machine is added. He was kind enough to show me his machine—the left hand operating the bellows and the right hand operating the pipes—imitating the various sounds of the human voice.

The research of Kratzenstein is found in the *Observations sur la Physique* by Rozier, 1782 supplement, p. 758. He also built a machine imitating the vowels, which consists of different ratios of the exterior and interior openings. I take this occasion to note that there are ten possible vowels. The vowel *a* is formed by leaving open all of the outside and the inside parts of the mouth. There are three series of these vowels that can be counted:

1. If the outside stays open and the inside narrows a bit:

   a

   ò (open *o*, as in some English words, and *aa* in Danish and *å* in Swedish)

   ó (ordinary *o*, which is also called a closed *o*)

   ou (which is expressed in Italian, in Spanish, in German, etc. by *u*; in Dutch by *oe*)

2. If the outside stays open and the inside narrows a bit[1]:

   a

   è (open *e*, which is expressed in French as *ai*, and in German by *ä*)

   é (closed *e*)

   i

3. Where the outside and the inside are narrowed at the same time:

   a

   eù (open, as in the word *bonheur*, in between *ò* end *è*)

(continued)

---

[1] In the original text, #1 and #2 are exactly the same, although, from the context, it appears that the author intended to say: "If the inside stays open and the outside narrows a bit."—*MAB*

éu (closed, as in the word *affreux*, or as ö in German, Danish, and Swedish, and as *eu* in Dutch, in between *ó* and *é*)

u (which is expressed in German as *ü*, in Danish and Swedish as *y*, and in Dutch (as in French) by *u*; in between *ou* and *i*)

To view these all at one time, they can be arranged in the following way:

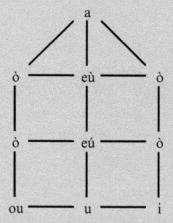

The vowels are not pronounced immediately one after another, without lightly touching the intermediates. There are as many diphthongs as there are possible ways to pronounce two vowels in a syllable.

53. Differences in Pipes

In organ pipes and other wind instruments, the airstream is the sounding body which makes longitudinal vibrations. One easily sees that the instrument itself is not the sounding body because the sound is not changed by the material with which it is constructed, the thickness of the walls, the diameter, the different ways in which one squeezes it in any location, or however any sound is prolonged. If, for example, pipes of the same shape are made of wood, metal, glass, etc. the differences in timbre seem dependent on the differences in the air friction against the walls or on weak resonance of the same walls. If one simply blows in a pipe, there is no focus to the sound; for that would produce only a progressive motion of the air, which is not a sound. It is necessary that the air enters through a narrow slit, or that it vibrates an elastic blade, where the vibrations produced are similar to the vibrations in the continuous airstream in the pipe. Or, at least, that a thin airstream is pressed with force, breaks itself against the splitting edge of an angular body, and passes almost in the direction of the axis in front of the end of the airstream.

The tone is dependent on:

1. The manner in which the air is blown
2. The width of the airstream contained in the pipe

If one of these two causes is predominant, it is sufficient to determine the sound. But if there is not such a preponderance, there is not an exact focus of the sound unless both of the two operate at least close to the same effect, because each cause tends to produce another sound.

54. Reed Stops

In the type of organ pipe called a *reed stop*, the sound depends principally on the manner in which the air is blown. The entering air vibrates a thin blade of copper, called a *tongue*, pressed toward the reed by a metal wire which is called the *regulating wire*, the top end of which has a notch or a hook to tune the pipes by pushing the regulating wire up or down by a tuning hammer. This increases or diminishes the vibrating part of the tongue, and at the same time enlarges or narrows the slit through which the air current enters.

The part of the pipe where the air vibrates longitudinally is longer for the graver sounds than for the more acute sounds, but not as long as those in other organ pipes, because the vibrations of the tongue force the contained air in the pipe to vibrate at the same time, which is against its nature. For this reason, the sound of these organ pipes is harsher than others; but in adding the other softer organ pipes, they serve to increase the power, especially to the graver sounds.

55. Flue Pipes

In organ pipes that are called *labial pipes* or *flue pipes*, as also in other wind instruments, the frequency of the vibrations is dependent mainly on the length of the airstream, in a way that one cannot produce sounds other than these, which are the inverse of the length of the vibrating parts of the air.

There is also always something that can be regarded as a reed, but it must be more within reach of the pipe than the reed stops mentioned previously. The types of organ pipes of which I speak have a transverse blade intersected in sloping manner, known as a *bevel*, which the air strikes perpendicularly in the way that air blown through the mouth can go out only through a narrow opening. This air going out, in the shape of a thin blade, hits the edge of the upper lip of an opening or slit called the *lumière*, and puts in motion the air contained in the pipe.

The *oboes*, the *bassoons*, etc. also have a reed that consists of two blades between which the air is forcefully pushed, shaking the sharp edge. In the *trumpet* and the *hunting horn*, the levers that are squeezed more or less firmly function as a reed; in the *flute*, the levers have the same function; the blade of air hits against the angular edge of the opening. In the *chalumeau*[2] there is a type of tongue, etc.

To produce the different sounds to which the same wind instrument under the same circumstances is predisposed, it is necessary to squeeze the levers more or less firmly and push the wind with more or less force. The small-diameter pipes, because of their length, will return acute sounds more easily than those pipes with larger diameters, where the airstream is divided into several vibrating parts. If there are lateral holes, the vibrating airstream is shortened when they are open, which raises the sound. The effects of these lateral openings still have not been successfully calculated, but the better experiments are those of Lambert in the *Mém. de l'Acad. de Berlin*, 1775.

Giordano Riccati (*delle corde ovvero fibre elastiche, schediasma* vii, par. 13) shows very well the passage of the vibrations in the air that depend on the mouthpiece, or on the manner of blowing, as those that are mostly determined by the length of the airstream. The reed of an oboe, inflated separately, produced grave and acute sounds differing by a sixth of an octave or more. But if the same reed was used in an oboe, and all of the holes were left open, the biggest difference in the sounds was nearly a fourth, and the sound was less defined, because the vibrations were in a longer airstream, lessening the effect of the manner of blowing.

If all of the lateral holes were closed, the vibrating airstream was too long to regulate its motion with ease by different ways of inflation; the biggest difference was thus nearly a tone; the intonation was false and very disagreeable because the sound of the mouthpiece was much too different from the one that suited the longer airstream contained in the pipe.

One sees therefore that each of these two causes (inflation and the length of the pipe) has its sphere of activity, where one seconds the effect of the other, as long as they do not go out of the boundaries of this sphere. As they compete to produce the same effect, their sphere of activity is extended; in both cases, the sound is formed more easily.

56. Kinds of Pipes

The modes of vibration and the series sounds are different if an organ *pipe* is *closed on one side* or if the *two ends are open*. It is always necessary to look at the end where one blows as open, even if it is put directly in the mouth, as in the horn and

---

[2] A predecessor and close relative of the clarinet. The lower register of the modern clarinet is often called the *chalumeau* register.—*TDR*

the trumpet.[3] The laws of vibrations are exactly the same as those of the longitudinal vibrations of rods (Sect. 5B). If one of the ends of a pipe is closed, the air vibrates as a rod with one end fixed; if the two ends are open, the air vibrates as a rod with two free ends; and if there is a way to put the air in motion in a pipe with the *two ends closed* (which would be done by blowing through a gap in the middle, as with a flute), the vibrations of the air would be similar to those of a rod fixed at both ends, or to the longitudinal vibrations of a string.

57. Explanation of the Manner in Which Vibrations Are Made

In all modes of vibrations, the compressions and expansions of the air are formed alternately, such that each portion of the air alternately approaches and recedes from the nodes of vibration. These small compressions and expansions, as also the longitudinal oscillations of the air molecules, are quite unequal in different places. At the nodes of vibration, the compressions and expansions are maximum (because the actions of all the other parts of the air contribute to this effect), but the oscillations are zero; the further a part is removed from a node of vibration, the more the compressions and expansions diminish, while the oscillations of the molecules increase; and at the middle between two nodes, or at an open end, the oscillations are maximum, but the compressions and expansions are zero, and the air density always remains the same as that of the open air that surrounds the pipe.

58. The Difference Between Simple and Double Parts

If the airstream contained in a pipe separates into any number of vibrating parts, the length of the part located at an open end is always one half that of the part contained between two nodes of vibration, so that the latter can be regarded as composed of two parts of half its length, which will be contiguous with the free end. I will therefore, in order to facilitate the demonstrations, call the part between two fixed limits, a *double part*, and the part located at an open end, or one half of a part contained between two fixed limits, a *simple part*. In such a simple part, the maximum compressions and expansions, without oscillations of the molecules, take place at one of the ends, while at the other end maximum oscillations occur, but with no compression or rarefaction.

---

[3] The mouthpiece end of a horn or trumpet acts as a closed end of the pipe. The acoustical behavior of the mouthpiece and the pipe are designed to make it play even as well as odd harmonics.—*TDR*

59. Simplest Motion of Air in a Closed Pipe

The *simplest motion* of air contained in a pipe, one end of which is closed, is that in which there is only one *simple part*. The air alternately approaches and recedes from the closed end (Figs. 17a, b), which has the same function as a node of vibration in other modes of vibration. This motion, which produces the deepest sound of which a pipe of the same length is capable, should be regarded as unity; both for the dimensions and number of vibrating parts, as well as for the number of vibrations that can be executed in the same time interval.

60. Simplest Motion of Air in an Open Pipe

When the *two ends* of a pipe are open, a node of vibrations is formed in the middle of the pipe for the simplest motion of the air. In this case, the *two simple parts* mutually approach and recede (Figs. 14a, b).

One will therefore have, as it were, two equal and closed pipes, where the layer of air in the middle, against which the other layers press from one side or the other, serves the function of a fixed separation. The sound is therefore an octave higher than the fundamental sound of a closed pipe of the same length, or the same as that of a closed pipe of half the length, but, because there are two of them, the sound is much stronger and more agreeable.

61. Other Motions of Air in Closed and Open Pipes

Aside from the simplest modes of vibration, still others can be formed if we change the mouthpiece and the force of the air, and especially if the diameter of the pipe is small in comparison with its length.

In the *second sound of a closed pipe*, a node of vibration is formed at a distance of one third the length from the open end where the air is blown in, and two-thirds of the length from the closed end, and the air is separated into a double part and a simple part, whose air layers mutually approach and recede, as in Figs. 18a, b.

The airstreams should be regarded as divided into three simple parts. The ratio of the frequency of vibration to that of the fundamental sound is as 3 to 1, the sound is therefore more acute by a twelfth or a fifth plus an octave.

In the *second sound of an open pipe*, there are two nodes of vibration, removed from the ends by a quarter of the length, and the airstream is divided into a double part in the middle and two simple parts at the ends, which is equivalent to four simple parts; the divisions and the reciprocal motions are represented in Figs. 15a, b.

The sound is to the first of the same pipe (Fig. 14) as 4 to 2, or more acute by an octave.

In the *third sound of a closed pipe* (Figs. 19a, b), there are two double parts and one simple, which is equal to five simple parts; the sound is to the fundamental sound (Fig. 17) as 5 to 1; it is therefore more acute by two octaves and a third and differs from the second sound (Fig. 18) by a major sixth or 5 to 3.

In the *third sound of an open pipe* (Figs. 16a, b), two double parts are formed at the middle and two simple parts at the ends, which is equivalent to six simple parts; the sound is to the first sound (Fig. 14) as 6 to 2, or more acute by a twelfth, and to the second sound (Fig. 15) as 6 to 4, or more acute by a fifth.

These explanations and Figs. 17, 14, 18, 15, 19, and 16, which represent the alternate motions, suffice to develop the idea of other modes of vibration, where a pipe, one end of which is closed, always divides into an odd number, and a pipe whose two ends are closed, into an even number, of simple parts. One will see also that the sounds are always in proportion to the numbers (or to the inverse lengths) of these parts. Consequently, if we regard *do* as the lowest of the piano (which I will express by *do 1* in accord with Par. 29) as the fundamental, all the sounds that can be produced on the same pipe, or on pipes of the same length, according to whether the end opposite that where air is blown in is closed or open, will be as shown in this chart:

Number of simple vibrating parts	1	2	3	4	5	6	7	8	9	10
Sound of a closed pipe	*do 1*		*sol 2*		*mi 3*		*si^b 3−*		*re 4*	
Sound of an open pipe		*do 2*		*do 3*		*sol 3*		*do 4*		*mi 4*

etc.

62. Ratios of Sounds Equal to the Natural Series of Numbers

The wind instruments normally used follow the same laws as those of an organ pipe whose two ends are open. Now, if one considers their sounds separately, without regard for the sounds of closed pipes, one would be able to change the series of sounds 2, 4, 6, 8, etc. into 1, 2, 3, 4, etc. by dividing by 2, and go an octave lower; one would therefore have the ordinary series of sound for the bugle, the trumpet, etc. which is the same for the sounds of the aliquot parts of a string. Par. 20 contains several remarks on the uses of these sounds.

63. The Shape of the Pipe Is Unimportant

It does not matter that a pipe of an organ or another instrument has an airstream that is straight or that bends because the air exercises the same elasticity in all possible directions. The series of sounds just mentioned is consistent within an instrument that is converging or diverging in any direction whatever. If a diverging pipe, a pipe whose diameter is everywhere the same, and a converging pipe have the same length, the sounds of the diverging one are slightly more acute, and those of the converging one slightly graver than the sounds of the pipe whose diameter is uniform. A pipe whose end is partially closed, as those that are called *chimney pipes*, should be placed between the open ones, in its effect; in closing the opening to a greater or lesser degree, one will have at one's disposal all the sounds between the deepest sound of the closed pipe and an octave higher than an open pipe. In this way, horn players make the sounds deeper by thrusting the hand into the mouth-piece, in order to produce sounds that are not included in the series mentioned, and that the instrument fails to achieve naturally. But this lowering is more limited in the horn because, by virtue of its shape, the opening can be closed only by thrusting the hand far enough forward, which by shortening the extent of the vibrating air, reduces the effect of closing, producing a contrary effect. Nevertheless, in order to harmonize open organ pipes, one folds the edge back slightly outwards or inwards, in order to raise or lower the sound. Thus, in all organ pipes, closed and open, the end where they are blown into is only open through the slit called the *lumière* which makes the sound slightly graver than if it had been a full opening; but the difference is less in long pipes than in those that are short.

64. Laws of Sound

The sound of pipes, if the mode of vibration is the same, is dependent on the length, the density, and the elasticity of the fluid that fills them. If n is the number of vibrations characteristic of each mode of motion, L the length of the vibrating airstream, G the weight, P the elasticity, which is equal to the air pressure, and h the height through which a body falls in 1 s, the number of vibrations that are made per second will be $S = n\sqrt{\frac{2hP}{LG}}$. The atmospheric pressure can be determined by the height of the mercury in a barometer. If the specific gravity of mercury is to that of the air as m is to k, and if a expresses the height of the mercury in the barometer, $\frac{P}{G}$ will be $= \frac{ma}{KL}$, and we will have $S = n\sqrt{\frac{2hma}{KL^2}}$ or $S = \frac{n}{L}\sqrt{\frac{2hma}{K}}$.

It then follows that:

The sounds of the pipe are inverse to the length if the other circumstances are the same.

The diameter of a pipe does not determine the sound, but in a pipe of large dimensions, the sound can be produced with great intensity.

On mountains of great height, the sound of a pipe will be the same as on the surface of the sea, because P and G increase or decrease together, always preserving this same ratio.

The frequency of the vibrations can only be changed by changing the ratio between the elasticity of air and its density. If the air has another specific gravity due to a mixture of other types of gas, or to variations in heat and cold, the pressure of the atmosphere remaining the same, its ratio to the weight, or $\frac{P}{G}$ (which can be called the specific elasticity) will be changed. Consequently, a pipe will give a more acute sound when it is heated than when it is made cold: in the climates of northern Europe, the extremes can amount to almost a tone. A wind instrument will never remain in tune with a stringed instrument in variations of temperature because the effect that the cold and heat have on the one is completely opposite to what it has on the other.

Experiments only establish these determinations of sound approximately, since the frequency of the vibrations found by experiment always exceeds what the theory gives us. The laws for the vibrations of the air in the pipe being the same as for the propagation of sound in open air, one will also find the number of vibrations for the first sound of an open pipe (Par. 60) by dividing the actual velocity of propagation by the length of the pipe. More information on this can be found in Part III.

Experiments made by Sarti on 19 Oct 1796, and communicated to the Academy at St. Petersburg, have shown that in a closed pipe, the length of which is 5 ft, there are (100 double vibrations or) 200 simple vibrations per second, which is in accord with the determinations of the number of vibrations given here.

65. Authors Consulted

The best research on the theory of wind instruments is found in the following dissertations:

Daniel Bernoulli on the sound and tone of organ pipes, in the *Mém. de l'Acad. de Paris*, 1762.

Observations of flutes, by Lambert (*Mém. de l'Acad. de Berlin*, 1775).

Leonhard Euler, *de motu aëris in tubis*, *Nov. Comment. Acad. Petrop.*, vol. XVI.

Research on the nature and the propagation of sound by Lagrange, in *Mélanges de Philosphie et de Mathématiques de la Société de Turin*, Vols. 1 and 2.

Giordano Riccati, *delle corde ovvero fibre elastiche*, *schediasma* 5, 6, 7.

66. Sound Produced by the Combustion of Hydrogen Gas in a Tube

The sound produced by hydrogen gas in a tube does not differ from the sound of wind instruments. The tube is not the sounding body, for the same reasons that the wind instrument is not one. In order to produce such a sound, one develops hydrogen gas by well-known means, in a closed bottle, from which the gas flows out through the tube of a thermometer or of a barometer that penetrates the stopper. The emerging gas is ignited (with the necessary precautions); a tube of glass or metal is then held over this flame. This tube may be open or closed, of arbitrary diameter and length, or a bottle, a horn, or other similar vessel. It is held over the flame in such a way that the flame will be forced up a certain distance from the opening; the sound is ordinarily very similar to that of a harmonica but sometimes much stronger. The flame should be small and tranquil. It should be made smaller as soon as the sound can be heard. In order that the flame be thus disposed, and to avoid the result that the tube, through which the gas emerges, is not closed by condensed water vapor, it is convenient to make use of a tube slightly larger than that of a barometer, of which one has reduced the upper opening with a valve.

The laws of vibration are the same for these sounds as for those of the organ pipe. The flow of hydrogen gas, the flame, and perhaps also the current of atmospheric air entering from below, in order to fill the vacuum caused by the absorption of oxygen gas, all contribute to produce vibrations in the air contained in the tube or vessel, in the direction of its length. These vibrations make themselves heard very loud if a finger is held under the lower opening of the tube.[4] If the upper end of the tube is closed, the sound is an octave lower than if the same tube is open at both ends: one can therefore raise or lower the sound by closing (more or less) one of the openings by one's finger or in another way. The sound is the same even if one blows into the opening, it is inversely proportional to the length of the tube, but it does not depend on the diameter. I have succeeded sometimes in the production of a second sound and even the third in a sufficiently long and narrow tube, by not thrusting the flame as much; the possible series of sounds is then, as in organ pipes, an odd number in a closed tube and an even number in an open one.

67. Sounds of Different Kinds of Gases

The frequency of vibration of different gaseous materials, when the elasticity caused by the pressure of the atmosphere is the same, will, according to theory, be as the square root of their specific gravity. Here are the results of several experiments that I performed in Vienna, with Professor von Jacquin, on the sounds of different types of gas, with which the same organ pipe was filled, surrounded, and blown through.

An open steel organ pipe, where the length of the vibrating airstream was about 15 cm, was fixed in the neck of a glass bell, furnished with a stopcock and a bladder attached from outside. After having emptied the bladder of air, and filled the bell and the pipe with water, by plunging under the water, a quantity of gas soon entered the bell and the bladder, so that the height of the water that closed the bell was the

---

[4] The relative positions of the tube opening, the flame, and the finger are not clear from Chladni's description.—*TDR*

same inside and out; the compression of the gas was, therefore, the same as that of the open air.

The pipe was inflated, with a lot of care, by a very light compression of the bladder, to avoid any change in the sound. The temperature remained the same for the whole experiment, about 10–12° Reaumur.[5]

At first, to know if the frequency of the vibrations in a gaseous material was changed by this closure, we filled up this apparatus with air from the atmosphere: the sound was the same when the air was free, but weaker.

The sound from using oxygen gas was graver by a semi-tone or nearly a tone, which is somewhat in accordance with the theory.

The use of nitrogen gas did not conform itself to the theory. It was presumed that the oxygen gas, weighing more, should vibrate a little more slowly, and the nitrogen gas, being lighter, would vibrate a little quicker than the atmospheric air. And, that the sound of the atmospheric air should be an average between the sounds of the two types of gas of which it is composed. However, the sound of the nitrogen gas (produced in three different ways) was always slightly graver than the sound of the atmospheric air, by almost a semi-tone.

To see if the gas used was lighter than the atmospheric air, we weighed one of these three types. The quantity contained in a glass sphere weighed 17 grains, and the same quantity of atmospheric air weighed 18 grains.

A mixture of nitrogen gas and oxygen gas produced a sound slightly more acute than that of one of these fluids; it was equal to that of atmospheric air. But before the mix of these two fluids became homogeneous, by repeated pressure on the bladder, the sound was not perceptible, because the vibrations could not become isochronal.

The hydrogen gas produced much more acute sounds than the atmospheric air, but not by as much as demanded by the theory. The sound of the gas produced by iron and sulfuric acid was more acute by slightly more than an octave; by zinc and muriatic acid by a ninth; by the vapors produced by passing through a heated iron pipe, by slightly more than a minor tenth.

The sound of carbonic acid gas was more grave by about a major third than that of the atmospheric air, which conforms to the theory.

The sound produced by the nitrous gas was hardly significant; as much as it was possible to observe, it was slightly more grave by a semi-tone than that of the atmospheric air.

These imperfect experiments, which it would be necessary to repeat with more exactitude, show at least that the lighter gases vibrated faster than the heavier gases, except where small differences were caused by chemical attributes.

---

[5] Reaumur temperature scale—equal to about 55–60 °F, and about 12–16 °C.—*MAB*

Section 5: Vibrations of a Rod or a Straight Strip

A. Transverse Vibrations

68. Different Cases

Transverse vibrations of a rod or a straight strip (that is to say a rigid body, straight, thread-like, where the changes in shape can be expressed by bent lines) are different, according to whether one or two ends are fixed (in a vise or in a wall), or supported (by a motionless body), or free. Here are the possible cases in which there is a difference in the changes in shape or the ratios of the sounds on which it depends:

1. If one end is fixed, and the other free
2. If one end is supported, and the other free
3. If the two ends are free
4. If the two ends are supported
5. If the two ends are fixed
6. If one of the ends is fixed and the other supported

So as not to be misheard, it is necessary to notice that this is only a question of vibrations of cylindrical rods or prisms and straight strips (or parallelepiped rods), that are not susceptible to other transverse vibrations, and that can be described by a curved line. Strips or wider blades pertain to rectangular plates, the vibrations of which will be explained in Sect. 7.

To create experiments, one can use rods made of glass, iron, or other rigid materials. If one uses straight blades, the vibration nodes will be visible by the same means as that for the vibrations of the plates.

© Springer International Publishing Switzerland 2015
E.F.F. Chladni, R.T. Beyer, *Treatise on Acoustics*,
DOI 10.1007/978-3-319-20361-4_8

69–74. Vibrations of a Rod, One or Both Ends of Which Are Fixed, Supported, or Free

69. In the first case, where *one of the ends is fixed and the other free*, the simplest way of vibrating is that where the entire rod vibrates (Fig. 20), alternately, first on one side and then on the other, and the axis is not intersected by the curve, but only touches the fixed end. It gives the gravest sound that can be produced on the same rod. In the other ways of vibrating, the axis is intersected by the curve one, two, three, or more times. The best means of producing these sounds as wished is to touch lightly a node of vibration with a finger, and to stroke a vibrating part with a violin bow. In the second sound (Fig. 21), the frequency is as that of the first, as the square of 5 to the square of 2, or as 25 to 4; the difference between the two sounds is therefore of two octaves and of an extreme fifth.

In separating the frequencies of the first sound from the frequencies of all the others, counting twice (Fig. 21), they will be between them as the squares of the numbers 3, 5, 7, 9, etc.; the third, or where there are two nodes, will therefore surpass the second by an octave and an extreme fourth; in the fourth, the pitch will increase by about an octave; in the fifth, by about a major sixth, etc. To decrease to the same pitch, all the ratios of the sounds to which the rod or strip is susceptible, in the case mentioned and in all others, I regard the sound, for the simplest motion (Fig. 20) as the *do* one octave below that of the first *do* of the piano, or, following the expression adopted in Par. 29 as *do*; thus, the possible ratios of such a rod will be:

Number of nodes	0	1	2	3	4	5
Tone	*do*	*sol*$^{\#}$ 2	*re* 4	*re* 5−	*si*$^{\flat}$ 5	*fa* 6+
Numbers whose squares correspond to these tones	(2)	(5) 3	5	7	9	11

etc.

The possible series of sounds, therefore, regarding the fundamental as unity, is $1, 6\frac{1}{4}, 17\frac{13}{36}, 34\frac{1}{36}, 56\frac{1}{3}$, etc., or expressed in whole numbers, 36, 225, 625, 1225, 2025, etc.

One uses the first tone of a similar rod on an iron violin. I used it myself for the tonometer described in the note for Par. 5.

70. In the second case, where *one of the ends is supported and the other free*, vibrations of the entire rod do not exist. In the modes of vibration in which there are nodes, they are slightly farther from the free end than in the first case, and the shapes into which the rod is bent are different, as are their corresponding tone ratios because a part with a fixed end is more hindered in its vibration than if the same end were supported. In the simplest vibration mode, a node of vibration is located approximately one third of the way along the rod from the free end; in the second, there are two nodes of vibration, and the one closest to the free end is slightly farther than one fifth of the length of the rod, etc. In order to produce these modes of vibration at will, one must, by holding lightly with the finger a point where there is a node, support the rod against a table or other stationary object and put it into motion by a violin bow, the middle of a vibrating part of a free end. The possible series of tones is equal to the squares of the numbers 5, 9, 13, 17, etc.; the gravest tone in this case is that which occurs in the first case, as 624 is to 144. The same rod or strip, which would have given the tone mentioned in the first case, will give the following tones in this case:

Number of nodes	1	2	3	4	5	6
Tone	*re* 2	*si*^{b} 3	*si* 4−	*sol*$^{\#}$ 5	*re*$^{\#}$ 6+	*la* 6
Numbers whose squares correspond to these tones	5	9	13	17	21	25

etc.

71. If, in the third case, *the two ends are free*, there are two nodes in the simplest mode of vibration (Fig. 24), in the second (Fig. 25) there are three, etc., and the length of the arc between two nodes is approximately twice that of a part located at one end. The deepest tone is that of the first case as 25 is to 4, the second case as 36 is to 25, and the series of tones is as the squares of 3, 5, 7, 9, etc. The same rod, whose tones are mentioned for the first and second cases, will give the following tone when the two ends are free:

Number of nodes	2	3	4	5	6	7
Tone	*sol*$^{\#}$ 2	*re* 4	*re* 5−	*si*^{b} 5	*fa* 6+	*si* 6−
Numbers whose squares correspond to these tones	3	5	7	9	11	13

etc.

These tones are the same as in the first case (except for the first tone) although the curves are quite different.

In order to perform experiments on this object, we could put the rod or strip in two points where there are nodes, on stands of a material that is a little soft (for example, of cork) and, while pressing lightly on the support with the fingers, we strike or stroke a vibrating part with a violin bow.

We make use of the first mode of vibration (Fig. 24) for chimes, where one strikes strips of glass, metal, or wood. A clavier is also used, for example, at Stuttgart (where the instrument manufacturer Hauk makes them very well), at Paris, London, etc.

72. If the *two ends are supported*, which is the fourth case, the rod bends along the same curves as those of a vibrating string; but the ratios of the tones are very different because they are not equal to the natural series of numbers 1, 2, 3, 4, etc., but to the squares of these numbers. In order to perform the experiments, we press plates or other things against the two ends of the rod, so that they cannot be displaced, and we stroke the piece of a vibrating part with a violin bow, touching, if necessary, a node of vibration so as to produce the tones of the various parts of a string. In the simplest motion (Fig. 1), which is equal to that of the fundamental tone of the string, the tone is the deepest tone of the first case (Fig. 20) as 25 is to 9, to that of the second case (Fig. 22) as 16 is to 25, and to that of the third case (Fig. 24) as 4 is to 9. The same rod that gives the tones mentioned will in this case yield the following tones:

Number of nodes	0	1	2	3	4	5
Tone	$fa^\# 1$	$fa^\# 3$	$sol^\# 4$	$fa^\# 5$	re 6	$sol^\# 6$
Numbers whose squares correspond to these tones	1	2	3	4	5	6

etc.

73. In the fifth case, where *the two ends are fixed*, (for example, in two vises), the rod vibrates either as a whole, or divided into two, three, four, or more parts. But the curves, of which one can get an idea by comparing Fig. 26 with Fig. 1, and the ratios of these sounds, differ from the preceding case. The sounds are the same as in the third case, where the two ends are free, in spite of the diversity of the curves. The same rod will therefore give the following tones:

Number of nodes	0	1	2	3	4	5
Tone	$sol^\# 2$	re 4	re 5−	$si^b 5$	fa 6+	si 6−
Numbers whose squares correspond to these tones	3	5	7	9	11	13

etc.

The results of the experiments will never be exact, for one will not be able to squeeze the ends of a rod into two vises without shortening it slightly. If one squeezes it too tightly, it is too hampered for the small expansions necessary because of the different size of the curve and the straight line. But in squeezing it less, the vibrations will sometimes conform to those described in Pars. 72 and 74.

74. In the sixth case, where *one of the ends is fixed and the other supported*, the rod or strip also vibrates either as a whole, or divided into two, three, four, or more parts. But the curves and the tones differ from those which take place in the two previous cases. For the first tone, the curve, which is not symmetric at its two ends, is represented in Fig. 27. The tones of all the modes of vibration are the same as in the second case, where one of the ends is supported and the other free, in spite of the diversity of the curves. The same rod will therefore give the following tones:

Number of nodes	0	1	2	3	4	5
Tone	re 2	$si^b 3+$	si 4−	$sol^\# 5−$	re 6+	16
Numbers whose squares correspond to these tones	5	9	13	17	21	25

etc.

To perform the experiments, we can set one end in a vise and have the other end supported by another vise or a machine, a plate, or something else rather immobile. We therefore reduce the tones by putting a vibrating part into motion as in the previous cases.

75. Laws of These Vibrations

If n expresses the relative number that corresponds to each mode of vibration of a rod, D its thickness, G the specific gravity, h the height through which a heavy body falls in one second, and S the number of vibrations which are carried out per second, then the frequency of the transverse vibrations of a rod or strip, as also the vibrations of all rigid bodies whose shape is the same, will be: $S = \frac{n^2 D}{L^2} \sqrt[2]{\frac{hR}{G}}$. Now, if the material of the rods and the mode of the vibrating is the same, $S = \frac{D}{L^2}$, the tones then will be the more acute the thicker the rod, and if the lengths are different, the tones will be the inverse square of the lengths.

The size has no effect on the tone. If a rectangular sheet is large enough to be regarded as a plate, the modes of vibration that correspond to those of a rod will then give the same tones as if the size were only that of a narrow strip but the force would be different.

The different sounds that can be produced on the same rod can be expressed by n^2, i.e., by the squares of certain numbers that mark the arithmetic progressions.

If the mode of vibration is the same, the stiffness of the material R will be $R = \frac{S^2 L^4 G}{D^2}$. We can therefore determine the sound by the stiffness of the material, which, if the dimensions of the rigid body are the same, will be $= S^2 G$, or as the squares of the number of vibrations multiplied by the weight of the material.

If the mode and the shape are the same, but the width is different, so that all the dimensions increase or diminish equally, the sounds, the mode of vibration being the same, will be as the inverse cube roots of the weights of the sounding body.

If some authors (for example, Nicomachus Gerasenus, Iamblichus, Gaudentius, Macrobius, Boethius) have claimed that Pythagoras had found the sounds of hammers in a forge, corresponding to their weights, this does not conform to nature; the sounds being rather as the inverse cubic roots of the weights. The same authors also maintain that Pythagoras had found the sound of strings in the ratios of the holding weight, which is not true at all; the sounds being as the square roots of the tension.

76. Authors Consulted

Daniel Bernoulli first successfully analyzed the transverse vibrations of rods and strings in Vol. 13 of *Nov. Comment. Academiae Petrop.*

L. Euler, after having published some very imperfect research in his *Methodus inveniendi lineas curvas maximi minimique proprietate gaudentes, add. 1 de curvis elasticis,* par. 282ff, has given a complete theory in his dissertation: *Investigatio motuum quibus laminae et virgae elasticae contremiscunt*, in *Act. Acad. Petrop. pro ann.* 1779, pt. 1, p. 103ff. All the results in this dissertation conform to experiments except what he has added on the vibrations of rings.

Giordano Riccati has also published very accurate research on the vibrations of a rod, both ends of which are free, in his dissertation: *Delle vibrazioni sonore dei cilindri,* which appeared in Vol. 1 of the *Memorie di matematica e fisica delle Societa Italiana.*

B. Longitudinal Vibrations

77. Explanations

In addition to the vibrations that we have just discussed, a rod or strip of a sufficient length is also susceptible to an infinity of other vibrations, in which the entire body, or its parts, according to the way in which it is divided, contracts and expands in the direction of its axis (or of its length) tending alternately toward one node of vibration or another or toward a fixed point. At the nodes of vibration, the compressions and expansions are the greatest but there are no oscillations of the molecules; in the middle between the nodes, and at a free end, the oscillations of the molecules are the greatest but there is no compression or expansion. The more remote a point is from a node, the greater the oscillations and the smaller the compressions and expansions. A vibrating part that is found at a free end is always one half of the length of the part contained between two immobile limits, which should be regarded as being composed of two parts contiguous to the free end. I will therefore (as in Sect. 4) call a part located between two immobile limits a *double part*; and half of such a part, or a part located at a free end, one of the ends of which is immobile and the other mobile, I will call a *simple part.*

The laws of these vibrations are exactly the same as those for the longitudinal vibrations of air in a pipe (Pars. 56–61); a rod, where one end is fixed and the other end is free, vibrates as the air in a pipe where the end is closed; if the two ends of a pipe are free, it vibrates as the air in a pipe where the two ends are closed; it makes vibrations as the air could vibrate in a pipe where the two ends are closed.

I published the first research on these vibrations in the *Act. Acad. Elect. Mogunt*, Erford, 1796.

78. Manner of Making the Experiments

To facilitate the experiments, it is necessary to use long, straight rods of a diameter that is not too big. It does not matter if the shape is cylindrical, prismatic, or flattened. If the surface is polished, it will be easier to produce the sounds. If the rod is metal or wood, it is necessary to rub a vibrating part in the direction of the length with a small piece of cloth, on which one puts a little rosin powder. But if the rod is made of glass, for example, if one is using the long tubes made for barometers and thermometers, it is better to put a little very fine sand or pumice on this cloth, moistened with water. In all cases when the rod is divided into vibrating parts, it is held by the fingertips at a point where there is a vibration node. Since these sounds are extremely acute, we must use very long rods.

79–81. Different Cases

79. If *one of the ends is fixed (in a vise) and the other free*, the entire rod can become alternately longer and shorter (Figs. 31a, b), in such a way that every molecule alternately approaches and moves away from the fixed end; there is therefore only a single vibrating part (according to Pars. 58 and 77). One must regard this motion as whole, both for the sound that is the deepest of all, as for the lengths and for the number of vibrating parts in the other longitudinal motions. In order to produce this sound, the rod may be rubbed along its entire length in the manner indicated in Par. 78. In the second sound (Figs. 32a, b), there is a node at a distance of one third of the length of the rod from its free end, which point the vibrating parts mutually approach and recede from, as the air in Fig. 18. The rod is divided therefore into a double part and a simple part, which is equivalent to three simple parts; the sound is to the first as 3 to 1, i.e., more acute by a twelfth or a fifth of an octave. This sound is produced by holding the node lightly with the fingertips, and rubbing the middle of the double part of the free end. In the third sound, the rod is divided into two double parts and one simple part, the motions of which are represented in Figs. 33a, b; the number of simple parts and the sound correspond to the number 5; the sound is therefore more acute than the second by a major sixth. In all the other longitudinal motions of a similar rod, the number of simple parts and the sound will be as the other odd numbers.

80. If the *two ends are free*, a node is formed in the middle in the simplest motion (Figs. 28a, b) which the simple vibrating parts approach and recede from while supporting one another. The sound conformable to the reciprocal of the length, or to the number of vibrating parts, is to the first sound in the previous case (Par. 79) as 2 to 1; it is therefore higher by an octave. To produce this sound, we must hold the rod at the center with the fingertips and rub one of the two halves. The second mode of vibration is that in which two nodes are formed, distant from the end by a quarter of the length of the rod; the rod is therefore divided into four simple parts, the motions of which are shown in Figs. 29a, b. The sound conformable to the number 4 is higher by an octave than the first sound. The third mode of vibration is that in which the rod is divided into two double parts and two simple (Figs. 30a, b), which is equal to six simple parts; the sound which is also conformable to number 6 surpasses the second by a fifth. All the other possible modes of vibration will be conformable to the numbers of pairs for the division into simple parts and for the sounds.

81. If the *two ends are fixed*, for example, in two vises, the entire rod in the first mode of vibration (Figs. 34a, b) has a motion that alternates between toward one or the other fixed point; in the second sound (Figs. 35a, b), it is divided into two double parts, which are supported alternately at the fixed ends and at the node that is found in the middle; in the third sound (Figs. 36a, b), it is divided into three double parts, etc. The number of simple pairs and the corresponding sound will be the same as in the preceding case (Par. 80) where the ends were free.

The longitudinal vibrations of a string (Pars. 43–45) can be regarded as the vibrations analogous to those of a rod the two ends of which are fixed. The sound does not depend on the tension because it is of too small an amount compared with the internal stiffness, i.e., the resistance to compression or expansion of the material.

82. Relative Frequencies of Vibrations of Different Materials

The sounds of rods, if their material and mode of vibration are the same, are proportional to the reciprocal of their length: the thickness does not determine the sound if it differs a great deal depending on the different materials. Having performed experiments on the relative frequencies of longitudinal vibrations of different material, I made use of rods where the strips were as long as possible; but I reduced the results to a rod of two Rhine feet in length [0.6276 m] and to the first motion, where the two ends are free (Fig. 28). The airstream in an open organ pipe of the same length yields the first *do* below, or, according to the mode of expression adopted here, *do* 3; but the sounds of all rigid bodies are much more acute. A rod of the same length of:

whalebone yields	. . . .	*la* 5
tin	. . . .	*si* 5
silver	. . . .	*re* 6
green wood ⎤	. . . .	*fa* 6
yew ⎦		

If the fibers of these woods were to be absolutely straight, the sounds would be slightly more acute.

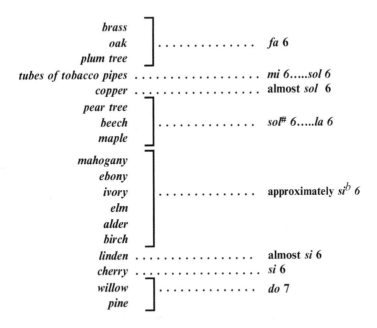

brass	
oak	fa 6
plum tree	
tubes of tobacco pipes	mi 6.....sol 6
copper	almost sol 6
pear tree	
beech	sol# 6.....la 6
maple	
mahogany	
ebony	
ivory	approximately si^b 6
elm	
alder	
birch	
linden	almost si 6
cherry	si 6
willow	do 7
pine	

If the fibers of these woods were not sufficiently straight, the sound would be graver, sometimes by a third.

glass	do# 7
iron or steel	
fir tree	do# 7 or almost re 7

However, these ratios of frequencies cannot be completely exact because of the internal differences of the same material, which can sometimes raise or lower the sound. All these frequencies surpass by a great deal that of air; the frequencies of glass, iron, and fir as much as 17–18 times as great. But if one excludes brass or tin, the sounds of which are very imperfect because of the low stiffness, the sounds of all the other rigid materials differ among themselves by almost an octave. It seems to me that the sounds of the different materials depend on the different ratios of the longitudinal stiffness, and on the specific density. If, for example, glass, iron, and fir have given out almost the same sound, we can presume that one of these attributes is compensated by another, in order to give almost the same result. It is reasonable that if n expresses the relative velocity consistent with the mode of vibration, L the length, C the stiffness, and G the specific gravity, the sound of a rod or strip will be

$$= \frac{n}{L}\sqrt{\frac{C}{G}}.$$

83. Laws of These Vibrations, Compared with Those of Transverse Vibrations

In order to distinguish more clearly the attributes and the entirely different laws of transverse and longitudinal vibrations of a rod or strip, I will put them next to one another in the following table.

Qualities of transverse vibrations	Qualities of longitudinal vibrations
Motion is produced in a transverse direction	Motion is produced along the length of the rod (or of its axis)
The rod forms different curved lines, while executing transverse incursions	The rod contracts and dilates in different ways along the axis
The sounds are in the ratios of the squares of certain numbers that make up the arithmetic progression	The sounds are in the ratios of a series of odd or even numbers
The sounds are as the squares of the reciprocals of the lengths	The sounds are as the reciprocals of their lengths
The sounds are in the ratio of the thickness	The thickness has no effect on the sound, except if they are greatly different, which can change the sound a little
They are as the square roots of the transverse rigidity, i.e., of the resistance to flexures, and, as the square roots of the reciprocals of the specific density	It is reasonable that the sounds are as the square roots of the longitudinal rigidity, i.e., of the resistance to the compressions and dilations, and, as the square roots of the reciprocals of the specific density

C. Torsional Vibrations of a Rod

84. Explanation of These Vibrations

A rod or strip is still susceptible to other forms of sonic vibrations, which I will call *torsional vibrations*, because the rod or its parts (separated by the vibration nodes) rotate around the rod, alternately in opposite directions, through extremely small distances, in such a way that one part turns to the right, while the part beyond the node turns to the left. At each point, these torsions are much smaller when this point is closer to a node; to the same nodes there is no point of motion.

I published the first research on these vibrations in Vol. 2 of the *New Memoirs of the Society of Friends of Natural Science in Berlin* (*Neue Schriften der Berliner Gesellschaft natur-forschender Freunde*) 1799.

85. Manner of Their Production

These vibrations are most easily produced on glass cylinders, the surface of which is polished, if one works in the same way as I have indicated for the longitudinal vibrations in Par. 78, except that it is not necessary to rub longitudinally but around the axis, in a circular direction, to the right or to the left. Sometimes, these can be produced on prismatic or parallelepiped rods by rubbing very lightly with the fingertips, or with a violin bow, in a diagonal direction, while taking the necessary precautions to avoid transverse vibrations.

If we put a little sand on the surface of a prismatic or parallelepiped rod, it will remain at rest on a narrow longitudinal line in the middle of each side, and also at the nodes of vibration.

86. Laws

The divisions of a rod, either *fixed at one end* or *fixed at both ends*, and the series of sounds that are consistent with these modes of division, follow the same laws as in the longitudinal vibrations, except that *the sound of a cylindrical or prismatic rod is always lower by a fifth than that of a rod, divided in the same manner, vibrating longitudinally.*

87. Application of These Vibrations to Those of a Plate

These torsional vibrations seem to merit special attention because they may be able to furnish a means for determining by theory the vibrations of plates that cannot be expressed by linear curves. In these same torsional motions that are of concern here, some nodal lines will be shown on a larger rectangular strip or sheet, as in Figs. 49, 50, 54, 55, 56, 59, 60, 61. All the motions of plates, where there is a nodal line in the length direction, could be reduced to these torsional vibrations by regarding this line as the axis, as, for example, in Figs. 63, 66a, 74a, 99–102, 183–187, and many others where the same torsional motions are modified differently by the shape of the sounding body.

Section 6: Vibrations of a Bent Rod

88. Vibrations of Forks

The vibrations of a *fork*, that is, of a rod or strip bent in the middle in such a way that the two branches are parallel, do not differ essentially from the transverse vibrations of a straight rod whose two ends are free (Par. 71), and cannot be judged exactly without comparing one with the others. If one bends a straight rod or strip of iron, copper, glass, or other sufficiently sonorous material, so that it is folded approximately as is shown in Fig. 37 *aa*, *bb*, *cc*, *dd*, and *ee*, one can observe the passage of the motions and the sounds from a straight rod to those of a fork. By the bend in the middle, as in general for each flexure of a vibrating part, the nodes become closer and closer together, as I have indicated in Fig. 37 by the small dotted lines. Each sound becomes graver if there were the same number of nodes on a straight rod or strip; so that the series of sounds which are consistent with the motion of a similar rod, and which are equal to the squares of the numbers 3, 5, 7, 9, etc., pass into a different form.

In the simplest motion, the two branches approach and recede mutually from each other, and the fork executes alternately the forms represented in Fig. 38, *npgqf* and *bphqm*. In comparing Figs. 24 and 38, we find that they are not essentially different, but that the axis is changed and two nodes are brought sufficiently close together by the bending in the middle, in order to regard them, without particular attention, as a single node.

The sound is graver by a minor sixth (or rather, as it seems to me, by a superfine fifth, 16:25 or 4^2 to 5^2) than the first sound of the same rod if it were straight.

A fork is not susceptible to a mode of vibration where there would be three vibrations, one in the middle and one on each branch, conforming to the second type of vibration of a straight rod (Fig. 25). The more a rod is bent at its middle, as in Fig. 37, the more difficult it is to produce this mode of vibration and finally it becomes completely impossible.

The second type of vibration of a fork is that in which there are four vibration nodes (Fig. 39) *mnte*, two very close to the middle and one on each branch; the fork

© Springer International Publishing Switzerland 2015
E.F.F. Chladni, R.T. Beyer, *Treatise on Acoustics*,
DOI 10.1007/978-3-319-20361-4_9

bends alternately on the curves *pdhgc* and *kfqzb*, and the sound is more acute than the first, by two octaves and a superior fifth; the first sound being to the second as the square of 3 is to the square of 5, or as 4–25.

But the first sound must be regarded as isolated from the series of the others, which is, to count from the second, as the squares of 3, 4, 5, 6, etc.

In the third sound (Fig. 40), there are five nodes, one in the middle and two on each branch. The sound is more acute than the second by a minor seventh 9:16; in the fourth sound (Fig. 41), where the pitch is increased by an augmented fifth 16:25, there are six nodes; in the fifth (Fig. 42), there are seven and the pitch is increased by a diminished fifth 25:36, etc.

Here is the series of sounds of a fork made by bending the same rod whose transverse vibrations have given the sounds mentioned in the preceding section:

Number of nodes	2	3	4	5	6	7	8
	Fig. 38		Fig. 39	Fig. 40	Fig. 41	Fig. 42	
Tone	*do* 2	*missing*	*sol*$^{\#}$ 4	*fa*$^{\#}$ 5	*re* 6	*sol*$^{\#}$ 6	*re* 7–
Numbers whose squares correspond to these tones	(2)		(5) 3	5	5	6	7

etc.

This series of sounds, counting from the second, etc., is the same as if the rod were straight and supported at its two ends (Par. 72), counting from the third sound. In the modes of vibration where there are two nodes in the middle very close together (Figs. 38, 39, 41), the sounds are the same as those of a rod one end of which is fixed and the other free (Par. 69), but it becomes more acute by two octaves; because the equilibrium of the two branches, of which one rests against the other, makes them vibrate as rods of which one end is fixed.

In order to perform the experiments, it will be convenient to make use of parallelepiped rods or strips of several widths, where one rubs one end with a violin bow, touching the vibration node with the fingertips. Each node could be made visible by holding its point in a horizontal direction and applying a little sand.

Besides the transverse vibrations, a fork that is sufficiently long is also susceptible to torsional vibrations and perhaps also longitudinal vibrations.

89. Vibrations of Rings

A *ring*, that is, a cylindrical (or prismatic) rod bent into a circle and welded at its ends, is divided in its vibrations into 4, 6, 8, 10, or more equal parts, and the ratios of the sounds which correspond to these modes of vibrations are as the squares of 3, 5, 7, 9, etc. In order to produce each mode of vibration, the ring is placed on a small cork (or compressed paper or other material that is slightly soft) supported at three points where there are nodes, and pressed lightly with the fingertips on these supports, in order that the ring does not move; a vibrating part is then rubbed with the violin bow. The vibrations are more easily made visible if the rings are placed horizontally and are rubbed in a vertical direction, because the shape arched by the ring prevents its parts from vibrating with the same facility from inside and out. To rub the ring vertically, one can put the supports on a table in such a way that the vibrating part of the ring that one wishes to put in motion projects slightly beyond the edge of the table. For example, to produce the simplest motion, where the ring is divided into four vibrating parts, one places (Fig. 43) the ring near the edge of the table *ab* so that the two nodes *m* and *n* and one other (*p* or *q*) rest on the supports, and so that the part *mgn* projects beyond the edge of the table: the points *m* and *n* are then pressed a little against the supports with the fingertips; one then rubs around *g* from low to high. One can produce the other modes of vibrations in the same way if one changes the positions of the supports.

If the lowest sound from the ring is *do 2*, the following sounds can be heard:

Number of nodes	4	6	8	10	12	14
Tone	*do 2*	*fa#* 3	*fa#* 4−	*re#* 5−	*la* 5	*re* 6
Numbers whose squares correspond to these tones	3	5	7	9	11	13

etc.

A ring whose deepest sound is *fa#* 3, after having been disconnected and stretched into a straight line, will give in its transverse vibrations, the sounds mentioned in Pars. 69–74.

I published the first research on the vibrations of a ring in my first acoustic dissertation: *Entdeckungen über die Theorie des Klanges* (*Discoveries on the Theory of Sound*), Leipzig, 1787. The assertions of Euler (*de sono campanarum*, in *Nov. Comment. Acad. Petrop.*, vol. X, *and investigatio motuum*, etc. in the *Act. Acad. Petrop.*, 1779), and Golovin (in *Act. Acad. Petrop.*, 1781, p. 2) are not supported by experiment and application of the vibrations of a ring to those of a bell does not conform to nature.

If a ring is more extended in the diametral dimension, it should be regarded as a plate, and if more extended in the other dimension, it must be regarded as a pipe or cylindrical surface, and the theory of its vibrations would not be suitable here, but in the following sections, where it will be a question of the vibrations of a straight or curved surface.

90. Vibrations of Other Curved Rods

The vibrations of curved rods of other types, such as those of rods or strips of an unequal thickness or size, could be the subject of much research.

Section 7: Vibrations of Plates

A. General Remarks

91. Explanation

In most of the motions of a plate (as also in those of a bell and of a taut membrane), the changes in shape cannot be expressed by linear curves, as in the transverse vibrations of other sounding bodies, but by curved surfaces, differently in different directions; and the nodes are not the motionless points but motionless lines, which one can call *nodal lines*.

> My first research on the vibrations of plates are found in *Entdeckungen über die Theorie des Klanges* (*Discoveries on the Theory of Sound*), Leipzig, 1787.

92. Manner of Performing the Experiments

To produce every kind of vibratory motion of a plate, and to make visible the nodal lines, one must hold one (or more than one) point immobile, and put into motion a movable point, by means of a violin bow, after having spread a little sand on the surface. The grains of sand are repelled by the vibrations of the vibrating parts and accumulate on the nodal lines.

To this general rule, we must add several remarks in order to facilitate the experiments.

© Springer International Publishing Switzerland 2015
E.F.F. Chladni, R.T. Beyer, *Treatise on Acoustics*,
DOI 10.1007/978-3-319-20361-4_10

The plates can be of glass or of sufficiently sonorous metal, for example, pure copper or brass. One could even use plates of wood, but the figures are not regular because the elasticity is not the same in the different directions. Ordinarily, I used glass plates because it is easier to have them at one's disposal and because their transparency allows us to view the points where we touch them underneath. Very thin plates are preferable to thicker ones because they can be bent in different ways. The size is arbitrary; for simple figures, a diameter of 3–6 in. will be sufficient; to produce more complex figures, it is necessary to use plates that are much larger. In order that the figures appear to be sufficiently regular, it is necessary that the thickness be quite uniform. Mirror glass is not preferable to vitreous glass because the surfaces are not always parallel. It is necessary to remove the edge with a file or by rubbing with a piece of sandstone in order that the strands of the bow are not damaged.

It is necessary to hold the plate at a point where the two nodal lines intersect (if there are any), because if we wish to hold another point on a nodal line, it hinders too much the vibrations of the neighboring parts. Since these nodal lines are only mathematical lines, they have no thickness. For this reason, the figures in which there are no nodal lines that intersect, for example, Figs. 67a, 104, and 109a, are the most difficult to produce; those whose fingers are too thick, or who do not have sufficient strength, never succeed in it. One must hold the plate with the ends of the thumb and of another finger, and press with great force so that the plate does not move when the bow is applied. Those who do not have sufficient strength, or whose fingers are not capable of these experiments, can make use of the machine shown in Fig. 44, of which the lower part is attached to the table by a screw and the upper part serves to hold the plate at a suitable point, between the ends of a cylindrical piece of wood and a screw plated by cork, leather, or felt. There are also some figures which can rest on the edge of the point with the ends of the fingers against a fixed body as, for example, in Figs. 109b and 115.

A good look will be much more useful than any measurement in determining exactly the most suitable points where one should touch the plate in order to produce each figure because the thickness, the shape, and the makeup of the plate are almost never sufficiently exact for the figure to agree exactly with the measured form. The production of most very complex figures will often depend on chance; but to produce less complex figures, one must know in advance what one wants to produce and imagine each figure as if it were already visible. If a suitable point has not been touched with sufficient accuracy, so that the figure appears somewhat imperfect, the position of the fingers must be changed somewhat, in order to hit the correct point exactly. When an interesting figure has been produced by chance, and one wishes to reproduce it, the production can be facilitated by marking the points where the plate is held and where the bow is applied. Sometimes, the point where the plate is held and the point where it is put in motion belong to more than one figure. If one wants to produce a manner of vibration that is in motion with other sorts of vibrations, it is necessary to exclude other movements, touching a point with the extremity of another finger at a nodal line. If one presses a small piece of cork or another soft material to a suitable point on the plate with a fingertip, and touches another nodal line with another fingertip, it is the same as if one had clamped it in the manner that has just been described.

One must hold the bow firmly enough in a vertical direction and move it in a manner that it always rubs the same point of the plate, so that it does not move to the right or to the left. The bow must always be applied to the region of a vibrating part that is not too far removed from the point where one holds the plate. In those cases in which the same manner of treating the plate can produce several sounds, one must be careful to move the bow without stopping, with the speed and pressure that is most suitable for producing one of these sounds while excluding the other, for a motion or a different sound destroys that figure which one wanted to produce. In general, the simplest figures, where the sound is deepest, appear most easily if one moves the bow with more pressure and less speed. To facilitate more complex motions, where the sound is higher in pitch, it will be suitable to employ greater speed and less pressure. The figures are formed more distinctly if (the sound and the touch being the same) one uses the entire length of the bow and if finally, after reinforcing slightly the motion of the bow, one withdraws it suddenly, in order to allow the plate free resonance.

The sand that is sprinkled on the plate can be ordinary sand; any other similar material, for example, iron filings or those of another metal, will have the same effect. The figures are especially well expressed if the sand is not too fine because the fine particles attach too strongly to the surface. However, if a little fine powder is mixed with the sand, it can serve better to make us see also the centers of vibrations, that is, the points where the vibrating parts make the largest oscillations: the smallest particles of the powder accumulate at these points. A little sand must be put on the surface, and spread out evenly; if too much sand is found at one point and too little at another, one can make the sand go toward the place where it is lacking by holding the plate for several seconds in a direction inclined toward that side.

Among the persons to whom I have shown these experiments, there are almost always those who have formed wrong ideas that are difficult to correct. They imagine that they can produce on a plate such sound that they wish (as in shortening the string of a violin) and that each sound gives a certain figure. One must not say that such a sound gives such a figure, but that each figure has a certain ratio of sound to the others. One cannot produce such sounds as one wishes, but only all the divisions imaginable, or there can exist an equilibrium of the parts among themselves, and the sound of each figure (or type of division) is so much more acute as the vibrating particles are minute. Consequently, one cannot produce sounds with ratios very different from those which are found in music on the same plate; here it is not a matter of octaves, fifths, etc. The production of these sounds has no resemblance to the shortening of the string of a violin, but rather to the production of sounds where the string is divided into aliquot parts, and cannot give sounds other

(continued)

than those that correspond to certain numbers. This remark is only designed for those who have not understood the general laws of sounds given in Par. 32.

93. Several General Attributes of These Vibrations

Two vibrating parts separated by a nodal line always make their motions in opposite directions, such that one is above the rest position and the other below. The division of the plate into vibrating parts is always as regular as the shape, the fabric, and the more or less exact manner of producing the vibrations permit, because it is necessary that the parts, in order that they may vibrate at the same time, be in equilibrium with one another. A vibrating part, located at the edge of the point, is always smaller than a part enclosed between two nodal lines. These lines can traverse the plate in all sorts of directions, or return upon themselves, but they can never terminate at the edge of the plate. The shape of the nodal lines can resemble a hyperbola, a cycloid, an epicycloid, and many other curves, according to circumstances. Ordinarily, the curves of the two lines are serpentine or of two similar lines, separated by a straight line, mutually approaching and departing from one another.

Toward the points where the nodal lines cross, they always enlarge, such that the shape of the vibrating parts near these points is not angular but more or less rounded, often in the shape of a hyperbola. I have represented this enlargement in Figs. 63, 64, 66a, 69, 70, and 99–102. Those who wish to concern themselves with the geometry of these vibrations should not neglect either the rounding of the vibrating parts or the position of the centers of vibration, that is, those points where the oscillations are the greatest, and where the finest parts of the powder accumulate. These points are not found at the edge itself, but at a small distance from the edge; their shape is either round or drawn out lengthwise, following the figure of the vibrating parts.

The sounds of the figures where the interior of the plate is surrounded by nodal lines (for example, in Figs. 65, 67c, 68a, 104, 105, etc.) are distinguished from the others by a different timbre, being louder and less disagreeable. The effect is sometimes as if the sound were graver by an octave, which is really not the case.

94. Direction of the Nodal Lines

In all possible modes of vibrations of a plate, the figures of the nodal lines can be reduced to a certain number of lines, which either traverse the breadth of the plate, or which are parallel to the circumference or on parts of the circumference; for example, on a rectangular plate to a certain number of parallel lines, to one or another dimension; on a round plate, to a certain number of diametric and circular lines; on an elliptical plate, or a semi-ellipse, everything is stretched out, etc. So far as the size of the plate allows it, one can produce on each plate every manner of division that conforms to its shape (or every member of the progression of numbers on nodal lines). If several kinds of vibrations do not produce a regular figure, they will however be represented by the *distortion*s of the nodal lines which can be reduced to the primary figure.

95. Distortions of the Figures

These distortions of the nodal lines do not change the tone because each vibrating part preserves the same relative size. These are the first elements for assessing the nature of these distortions. Even the most complex figures can be reduced to the primary figure by imagining the same thing repeated more or less frequently.

Two lines or parts of straight lines that are intersected (Fig. 45c) can be separated in two different ways to form two curves (Figs. 45b, d); these curves can also be transformed into two parallel straight lines (Figs. 45a, e). Likewise, two lines or parts of parallel straight lines in one direction can become curved and when the curvature becomes stronger, pass by two straight lines that are cut, into some curves and finally into straight lines in the other direction. There is therefore no essential difference between the positions of the nodal lines represented in Figs. 45a–e. They must therefore be regarded as variations of the same mode of vibration that one can often produce at will by changing the location where the plate is touched.

Similarly, at the edge of a plate (Fig. 46 *mn*), both ends of lines that approach at an obtuse angle (a), or as part of a curved line (b), or as part of a straight line (c), are equivalent and interchangeable.

If the shape of a plate is not regular, or if the thickness is not the same everywhere, the figures will always be distorted.

96. Affinities of the Figures Among Themselves

The figures, even the most complex ones, have more relationship and affinity among themselves than one would have at first thought. If, after having produced the same figure on several plates of the same shape and size, one puts one close to another in a suitable manner, the nodal lines of one plate continue of those of the other, and the composition of several simple figures are composed from other more complex ones, which one can also produce on the larger plates. The compositions of four square planes, two semi-round, four triangular, etc. furnish examples.

97. General Laws of the Frequency of Vibrations

The sounds of plates, the shape, material, and mode of vibration being the same are proportional to the thickness and inversely proportional to the squares of the dimensions. If the materials are different, the sound will be directly proportional to the square roots of the stiffness and inversely as the square roots of the specific gravities.

98. Several Lessons for Those Who Wish to Be Concerned with the Theory of Oscillations

The present state of the most sublime analysis does not allow us a means of determining by theory the nature of these motions and to express them in terms of equations, except those in which the diameter of the sounding body bends in the same way and in which the nodal lines are like those in Figs. 47, 48, 51, 52, 55, 57, 59, 67a, 104, and 109a. We haven't proceeded very far in this research since the time of Euler, who expressed himself as follows (*Nov. Comment. Acad. Petrop. vol. xv, p.* 582):

   Quae adhuc de figura corporum flexibilium et elasticorum in medium sunt allata, non latius, quam ad fila simplicia sunt extendenda. Quin etiam omnia, quae in hoc genere sunt explorata, ad curvas tantum in eodem plano formatas sunt restringenda: quare longissime adhuc sumus remoti a theoria completa, cujus ope non solum superficierum, sed etiam corporum flexibilium figura definiri queat; atque haec theoria etiam nunc tantopere abscondita videtur, ut ne prima quidem ejus principia adhuc sint evoluta.[1]

   The assumption, of looking at a similar rigid membranous body as a network formed by curved lines steered in a direction that is applied or curved lines steered in another direction, does not conform to nature and will never give either results conforming to experiment or the appearance of explanation of several kinds of very

---

[1] The substance of Euler's statement, rendered in modern terms, is that the only vibrational patterns of two-dimensional plates susceptible to mathematical analysis are those that are mathematically equivalent to a one-dimensional rod or "thread" vibrating in a single plane. Viewing the figures referred to by Chladni in the preceding paragraph may help clarify this for the reader.—*JPC*

simple vibrations. Jacob Bernoulli did not succeed in making use of such an assumption in the *Nov. Act. Acad. Petrop.* 1787.

It seems to me that the only way to arrive at the theory of the motions will be a thoroughgoing examination of the torsional motions of a rod (Pars. 84–87). We must begin by giving a general formula for the torsional vibrations of a cylindrical or prismatic rod (whose two ends are free), and which presents in the first sound a node in the middle; in the second, two nodes extending one quarter of the length of the rod from its ends; in the third, three nodes, one in the middle and the others displaced by a sixth of the length from its ends, etc. In the extremely small motions in which each molecule makes around the axis, alternately to the right and to the left, and which are smaller as the molecule is less displaced from the node and from the axis (that is, of the longitudinal fiber at the middle of the cylinder), the motion of each longitudinal fiber, regarded separately, will be a curve which cannot be described on a plane but on a cylindrical surface. The frequency in the different types of motion will be as the reciprocals of the lengths of the vibrating parts. When one has succeeded in forming an exact theory of these motions of a glass cylinder or prism, it would be necessary to apply it to the same motions of a string, a band, or the large rectangular plate, which will give the type of vibrations where the nodal lines appear as in Figs. 49 and 50, or as in Figs. 63 and 66a, etc. More instruction on this can be found in Subdivision C of this section, which contains the research on the passage of vibrations of a square plate to those of a narrow strip. The second series of sounds, in which there is a nodal line in the longitudinal direction, is no different than the vibrations of a rotating rod. One can show:

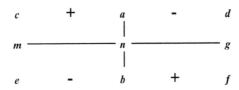

cdef as a rectangular blade, or as a part of a similar longer blade, where the lengthwise nodal line is intersected by the nodal line *ab*. Two vibrating sections separated by a nodal line always vibrate in opposite directions, in the way that the sections indicated here by + are above the ordinary position, while the sections marked by − are under, and the opposite will occur in the following vibration.

Now, if the section *angd* vibrates over, it is the same thing as if section *adfb* turned to the right. The same is true for the other half beyond the nodal line *abec*. When the sections *anmc* and *mnbe* vibrate in opposite directions, it is the same as if section *abec* turned to the left, supposing that the eye of the observer was at *g*. In the ensuing vibration, *adfb* will turn to the left, and *abec* to the right, and so on.

During these alternating motions, the nodal line in the direction of the length *mg* will replace the axis (or the longitudinal fiber in the middle) of a cylindrical or prismatic rod that makes torsional vibrations; and the nodal line of the width, *ab*, will replace the node of the same rod. A similar blade can be longer and will also be able to be similarly divided into several parts, resembling the preceding, where the vibrations go in opposite directions. The distances of one sound or another in a very narrow blade are as the natural series of numbers, larger if the blade is larger, in the manner of those in a square plate, as 2, 5, 10, etc. instead of 1, 2, 3, etc. The speed of the first sound will diminish as the width increases.

Perhaps, to determine the more complex vibrations, where there is more of a line in both directions, for example, in Figs. 78a, 71a, 75, etc. we could consider the plate as being composed of several similar contiguous sections on one side.

When the vibrations of a rectangular blade must be determined, the same principles can be applied to plates of another form. Thus, Figs. 99–102 will be the same for a round plate; Figs. 64 and 69 for a square plate regarded as a rhombus; Figs. 183–187 for elliptical plates, etc.

The combination of all the possible discoveries of this kind gives general expressions, by means of which one can predict the forms that the nodal lines must take on a plate of a given shape, shaking in a known manner.

One will never be able to advance much in the theory of these vibrations, until after determining exactly the nature of the distortions of the nodal lines (Par. 95), where, for example, Figs. 45a–e, can be regarded as equivalents; as also, Figs. 66a, b; Figs. 67a–c; Figs. 71a–c; Figs. 72a, b; Figs. 73a, b; etc. In these cases, the torsional vibrations and the transverse vibrations pass into one another without changing frequency.

It will be of some value to note that in a rod or very narrow band the sound of the first type of torsional vibrations is lower by a fifth (2:3) than that of the first type of longitudinal vibrations. But when the width is equal to the length (a square plate), the first type of torsional vibrations with the same ratio (2:3), represented by Fig. 63, is lower by a fifth than the first type of transversal vibrations, represented by Fig. 64. If the dimensions of a rectangular plate are as 2:3, the two types of ratios give the same sound.

B. Vibrations of Rectangular Plates in General

99. Different Cases

Rectangular plates are the first of which I set forth the vibrations because they can serve better than plates of other shape to show the passage of the vibrations of a narrow strip (Sect. 5A) to those of plates, which cannot be expressed by linear curves.

A rectangular sheet or plate (of glass, of sufficiently sonorous metal, etc.), in which the two dimensions are in any ratio, will be susceptible to different series of vibrations of a sound in the following cases.

1. If the ends are free
2. If one end is fixed and the other free
3. If both ends are fixed

We could also, as in the transverse vibrations of a rod, distinguish a fixed end, and a supported end, but if one or two ends are supported, one could scarcely produce any sufficiently regular vibrations and the figures would often resemble those of distortions of the figures of a free sheet or plate.

100. Vibrations of a Rectangular Plate Whose Two Ends Are Free

If the *ends* of a rectangular plate are *free*, the simplest series of vibrations is the same as that of a free rod (Par. 71) when it makes transverse vibrations. In the first kind of these vibrations, where the curvature of each fiber is as in Fig. 24, two nodal lines are formed, at a distance from the ends of almost a quarter of the length; these lines appear when one shakes sand upon the surface, as in Fig. 47. Figure 24 can be regarded as the profile and Fig. 47 as the plane of the plate. In the second mode of vibration, we must regard Fig. 25 as the profile, representing the curvature of each fiber, and Fig. 48 as the plane of the plate, representing the three nodal lines on which the scattered sand accumulates. The plate could also be divided into a rather large number of parts, where four, five, or more nodal lines are formed whose ends are removed from the ends by about half the length of a part contained between two nodal lines. Whatever the width of the plate, the series of sounds will always remain the same and will be equal to the squares of the numbers 3, 5, 7, 9, etc. To produce all these types of vibrations, we must apply the violin bow to one of the narrow edges, by clamping the node nearest this edge between the thumb and another finger. If one moves the finger slightly closer to the ends of the sheet, one will be able to produce all the series of these sounds. On a sufficiently wide plate, one could also produce, so far as the width permits, but with a great deal of difficulty, two or more nodal lines, according to the width, and the sounds would be between these same ratios as if the nodal lines were formed according to the width. But they will be more acute in proportion to the squares of the dimensions to which the position of the lines are related. All these nodal lines can also be curved in different manners, without changing the sound, as in all the other kinds of vibrations.

In addition to the modes of vibration, which are analogous to the transverse vibrations of a rod, the plate is susceptible to many others, which are not expressible in terms of linear curves. In these vibrations, nodal lines are shown at the same time in one direction, and in another at a right angle, or may also change position without changing the sound. In order to produce these vibrations, we must clamp a point where two nodal lines intersect and apply the bow to one of the longer sides near the end or in the middle of a vibrating part. The nodal line, according to its length, can be intersected by a line according to the width (Fig. 49), or by two, or by several.

One will be able to produce very easily all of this series by approaching the fingers closer and closer from the end and by keeping them always on the longitudinal line. The modes of vibration are the same as the torsional vibrations of a rod (Pars. 84–87, and 98); if the width is very small, the sounds correspond to the natural series of numbers; but, as the width increases, the distance of one sound from another also increases. The first sound changes in inverse proportion to the width; consequently, we cannot compare the first series of sounds made, because these sounds also depend on the width, which should have no influence on the sounds of the first series.

If the width of the plate permits it, we can also show two or more nodal lines, according to the length, intersected by the lines in the other direction.

Subdivision C of this section will contain more information and can be regarded as the continuation of this paragraph.

101. Vibrations of a Rectangular Plate with One End Fixed and the Other Free

A rectangular plate, *one end* of which is fixed and the other free, makes the simplest motions, like a rod in the transverse vibrations described in Par. 69. It can vibrate as a whole, in such a way that each fiber is bent along the curve shown in Fig. 20, and that there is no nodal line (Fig. 51); it can also form a node, separated from the free end by slightly less than a third (Fig. 52) or two nodes (Fig. 53), etc. The bow must be applied to the free end.

But, aside from these types of vibrations, which are similar to the transverse vibrations of a rod, there also exist others, where there are nodal lines in two directions. The first series, which presents a nodal line in the direction of the length, either alone (Fig. 54) or intersected by a line in the direction of the width (Fig. 55), or by two lines (Fig. 56), corresponds to the torsional vibrations of a rod, one of whose ends is fixed and the other free. If the width is less considerable, the sounds are like the odd numbers 1, 3, 5, 7, etc. The more the width increases, the greater the distance from one sound to the next increases. The first sound (Fig. 54) is lower by an octave than the first of the same sheet when the two ends are free (Fig. 49). If we want to compare these vibrations to those of a preceding paragraph; then the sheet, one side of which is fixed, will give those sounds that correspond to the odd number. If both edges are free, it will give those which correspond to the even numbers, and the sounds will be as the reciprocal of the lengths of the vibrating parts, regarding one part contained between two lines in the direction of the width, like the double of one part located at the free end. In order to produce this series of sounds, we must apply the bow to a suitable point on one of the longer sides and touch, for the first sound, the longitudinal line and for the others, a point where two lines intersect.

If the width permits it, we can also show two or more lines in the direction of the length, or one alone, or intersected by the lines in the direction of the width.

102. Vibrations of a Rectangular Plate with Two Ends Fixed

If the *two ends* of a rectangular plate are fixed, the simplest motions are the same as those of a pipe treated in the same manner. It can vibrate as a whole, or divided into two, three, or more parts (Figs. 57 and 58).

It is difficult to produce these because there is not a free end on which to apply the bow. However, one succeeds sometimes by a slow movement of the bow, applied on the side with more pressure, than for the other modes of vibration.

Besides these vibrations expressible by curved lines, the plate can vibrate again in many other ways (easier to produce), where there is a nodal line in the direction of the length, or alone (Fig. 59), or intersected by a line in the direction of the width (Fig. 60), or by two (Fig. 61), or by three (Fig. 62), etc. If the sheet is narrow, the sounds of these vibrations resemble rotating vibrations on a pipe with two fixed ends, corresponding to the natural series of numbers, and they are the same as if the two ends were free. One may easily produce these vibrations with the blade of a saw. I have represented the distortions of similar figures in Fig. 62b, c.

C. Vibrations of a Square Plate and Some Other Kinds of Rectangular Plates

103. Explanation

The vibrations of square plates will be explained first because these plates have the simplest ratios, the width being equal to the length. Then, by assuming one dimension as constant and the other as variable, I will show, in these plates of a gradually decreasing width, the passage to the vibrations of a narrow strip or rod which were described in Pars. 84–87.

I sometimes use the word *diameter* to express a direction parallel to one side because the word *dimension* would be too vague.

104. Nodal lines in One Direction or Another and Signs for Expressing Them

In all the modes of vibration of a plate, rectangular or square, the figures are always related to a certain number of nodal lines in one direction or another. Even if the lines are diagonal or twisted, they can always be reduced to a certain number of lines parallel to one side or the other.

To be more precise, I will express the lines in one direction or another by some numbers separated by a vertical line. Thus, for example, 3|0 will express the mode of vibration in which there are three lines in one direction and none in the other; 5|2 expresses the mode in which there are five lines parallel to one of the sides and two parallel to the other, etc.

105. Flexions of Nodal Lines

Nodal lines, which one can imagine as lines that are ordinarily straight, can be more or less curved. This *flexion* of these lines that are next to one another, or separated by a straight line, moves them closer together or farther apart (Par. 93). In some modes of vibration, the nodal lines are never straight lines on a square plate and in some other modes they are never straight on rectangular plates in some other ratios of the dimensions. For square plates, I indicate in the following table the number of flexions of nodal lines in one of the directions which often remain straight lines in the other direction. What I call a *flexion* here is the deviation of such a line toward one side, consisting of a departure and a return to the straight line which one can imagine as the ordinary form. The horizontal series of numbers (top row of the table) indicates the lines in one of the directions, and the vertical series to the left the lines in another direction:

	2	3	4	5	6	7
0	Figs. 64 and 65	Fig. 67	Figs. 72 and 73	Fig. 78	Fig. 85	
	1 flex	1 1/2 flex	2	2 1/2	3	3
1		Figs. 69 and 70	Fig. 74	Fig. 79	Fig. 86	
		1 flex	1 1/2	2	2	2
2			Figs. 76 and 77	Fig. 81	Figs. 87 and 88	
			1	1 1/2	2	2
3				Figs. 83 and 84	Fig. 89	Fig. 90
				1	2	2

106. The Essential Difference When the Nodal Lines Are Curved Inward and Outward

Several modes of vibrations, the number of nodal lines being the same, can appear in two entirely different ways, according to whether the flexions or part of the flexions of the exterior lines are going inward or outward. In the first case the sound is graver than in the second, with little exception. This difference can be noted in the figures in which there is an integral number of flexions, as in 2|0, 2|1, 4|0, 4|2, 5|3, 6|2, etc., but never in the case in which the number is $1\frac{1}{2}$, or $2\frac{1}{2}$, as in 3|0, 4|1, 5|0, 5|2, etc.

107. Types of Vibrations of a Square Plate

Among the different nodes in which a square plate can vibrate, the figures of the nodal lines are arranged in order, following the deepness or the elevation of the sounds, and I will explain them here in the same order.

Of all the types of vibration, 1|1 (Fig. 63) is the one that yields the deepest sound. We can produce it very easily by holding the plate in the middle and putting it into motion around a corner. The figure can often be changed into two curved diagonals *ehd* and *cmn*.

The mode of vibration that yields the next lowest tone is 2|0, where the two lines are curved from within (Fig. 64). The plate must be struck in the middle and a violin bow applied at one side. The sound is more acute by a fifth than the first. The figure can sometimes appear as the two curves, *cnd* and *emk*.

2|0, where the two lines are bent outward, ordinarily appears as a square with curved corners (Fig. 64) if, by gripping the plate near the edge, in the middle of a side, we put it into motion at the nearest corner. If the plate is held in the middle between two opposite sides between the tips of the thumb and another finger, the figure is drawn out in length and appears as two curved lines. The sound is more acute by a minor third than that of Fig. 64, and by almost a minor seventh than that of Fig. 63.

2|1 (Fig. 66a), where the sound is more acute by an octave and a major third than that of Fig. 63, appears very easily if, while holding the plate in a point where two lines intersect, we set it into motion in the middle of the right or left side. It is also sometimes possible, by small shifts of the fingers, to produce distortion into three diagonal curves (Fig. 66b).

3|0 is the type of vibration that is most suitable for easy showing of the distortions of the figures that are made without changing the sound. It can be produced at will on each square plate that is not too irregular as Fig. 67a, b, or c. One can also pass on three straight lines in one direction, through similar intermediate figures, to three straight lines in another direction, through similar intermediate figures, to three straight lines in another direction, by slight changes in the positions of the fingers that grip the plate. If one holds the plate at the point marked by *m* in Fig. 67a, and applies the violin bow at the point *n* in such a way that the point of contact and the point of rubbing are in the same diameter, three parallel lines will appear, and the motion of the plate will be exactly the same as that of a rod (Par. 71) in the second transverse sound (Fig. 25). For this effect, the plate should only be held by the tips of the fingers that are extended only a little, in order not to hinder the vibrations in the neighboring parts; one must move the bow very slowly and press more strongly than in producing the previous figures, or Fig. 67c. If the figures are very little advanced along the same diameter, without changing the point where motion has been produced, the lines curve as in Fig. 67b. If the fingers are advanced still further in the same direction, the curving of the nodal lines becomes stronger, and finally, these lines intersect in Fig. 67c. In the same way, by moving the fingers slightly toward the edge of the plate, one can convert Fig. 67c to Fig. 67b in one direction or the other. The sound does not change with these changes of the position of the lines, because each vibrating part always preserves the same magnitude, in order to vibrate with the same frequency.

2|2 (Fig. 68a) is very easy to produce if the plate is held at one of the points where two nodal lines intersect, and if motion is produced in the middle of one side, or at the corner closer to the point that has been touched. Figure 68b is a distortion of the same figure. I will not remark any further on the ratio of the sound for each figure because everything will be mentioned in the following paragraphs.

In 3|1, the exterior lines will never be straight on a square plate, but always curved toward the interior or toward the exterior. This must be regarded as an essential difference because the sound in the first case is graver than the other by slightly less than a tone. Ordinarily, 3|1, when the lines are curved toward the interior, appears as Fig. 69; the figure can also be similar to that of a round plate (Fig. 101b). To produce this mode (Fig. 70), one must hold the plate in the middle, and at the same time, touch the nodal line near the corner at which one applies the bow.

3|2 sometimes appears in the original form as in Fig. 71a, but more often as in Fig. 71b. If, while holding one point where two lines intersect, one rests the corner d or n on an immovable body, the zigzag $dpmqhn$ (Fig. 71a), then transforms to the right diagonal dn (Fig. 71b). The figure can also be changed into five diagonal curves (Fig. 71c) if one changes slightly the position of touching; and for this purpose it would be better to apply the bow near the closest corner.

4|0 can appear in two different modes, according to whether the exterior lines are curved twice, inward or outward. In the first case, the nodal lines appear as in Fig. 72a or 72b, in the second case, where the sound is more acute, as in Fig. 73a or 73b. I have never been able to produce four straight lines.

In 4|1, the lines can be straight, as in Fig. 74a, or be transformed into Fig. 74b, depending on the different points where one touches the plate and where one applies the bow. It is not difficult to produce these two different forms of the same mode of vibration on each regular plate. Sometimes, I have noted the passage of one to the other; the sound remains exactly the same.

3|3 ordinarily appears very regular, as in Fig. 75; but this figure can also be transformed into six curved diagonal lines of the same type as 2|1, 2|2, and 3|2.

4|2 has never appeared on a square plate in primary form; but Fig. 76 is a distortion of 4|2, where the exterior lines are curved inward. Figure 77 is a distortion of the same number of lines curved outward; the sound of Fig. 76 is slightly graver than that of Fig. 77. It is very easy to produce Fig. 76 if, while holding the plate in the middle, and touching at the same time (to avoid Fig. 64), with a fingertip, a point located on one of the curved lines, one applies the bow to the middle of one side. The best way to produce Fig. 77 is to grasp the plate in the middle of the two opposite sides with the thumb and another finger, while touching one of the little lines close to a corner with another fingertip, applying the bow to this angle.

5|1 rarely appears with straight lines; the commonest distortion is Fig. 79b; sometimes, Fig. 79a appears.

4|3 can be produced on sufficiently regular plates in the primary form (Fig. 80a), or changed into Fig. 80b or into seven diagonals (Fig. 80c).

5|2 appears regularly (Fig. 81a), or transformed into Fig. 81b.

4|4 appears sometimes regular as Fig. 82; it is also susceptible to the same distortions as I have noted in regard to 2|1, 2|2, 3|2, 4|3. Likewise, on larger plates, 5|4, 5|5, 5|6, 6|6, 7|6, etc. may come close (more or less) to 9, 10, 11, and a very large number of curves.

In 5|3, the lines were never straight. But Fig. 83 represents 5|3, where the lines are curved inwards, and Fig. 84, the same number of lines curved outwards. The sound of Fig. 84 is slightly more acute than that of Fig. 83.

6|0 has appeared very rarely regular, but is ordinarily transformed into Fig. 85.

In 6|0, the lines are very rarely straight; ordinarily this mode of vibration appears as in Fig. 86.

6|2 exists in two different modes: when the exterior nodal lines are curved inward and when they are curved outward; the difference of the sounds is almost a semi-tone. In the first case, the nodal lines appear as in Fig. 87a or b, transformed in other ways; in the second case, the nodal lines are as in Fig. 88a, or more often as in Fig. 88b.

6|3 can appear as in Fig. 89a, but it is much easier to produce Fig. 89b, which is only a distortion of the same figure. One must hold the plate at a point where two lines intersect and (in order to exclude Fig. 67c) touch one of the small circles very lightly with the fingertip while, at the same time, applying the bow to the middle of this semicircle in such a way that the two points should touch and the point of rubbing be on the same diameter. It seems to me that 6|3 could also appear in the manner that the ends are curved twice outward, but I have not seen it.

If the size of the plate permits, one can also push much further the production of different and more complex modes of vibration, of which Figs. 90–96 are examples. Figure 90 is a distortion of 7|3, Figs. 91 and 92 of 6|4, Figs. 93 and 94 of 8|4, and Figs. 95 and 96 of 8|6.

108. Ratios of the Sounds

The ratios of the sounds which are common among all modes of vibration of a square plate are contained in the following Table, where I look at 1|1, the simplest mode of vibration, which gives the gravest sound, as *sol 1*; the horizontal series of numbers (top row of the table) indicates the parallel nodal lines on one side and the vertical series to the left indicates parallel lines on the other side.

Each ratio, except 1|1, 2|2, 3|3, 4|4, etc. is found here two times, to best see the series of sounds in one look, and to compare the sounds of a square plate

	0	1	2	3	4	5	6
0			Fig. 64 *re 2* / Fig. 65 *mi.fa 2*	Fig. 67 *sol$^\#$ 3+*	Fig. 72 *sol$^\#$ 4* / Fig. 73 *sol$^\#$ 4+*	Fig. 78 *fa 5−*	Fig. 85 *do 6−*
1		Fig. 63 *sol 1*	Fig. 66 *si 2*	Fig. 69 *si 3* / Fig. 70 *do$^\#$ 4*	Fig. 74 *si$^\flat$ 4−*	Fig. 79 *fa$^\#$ 5−*	Fig. 86 *do 2*
2	Fig. 64 *re 2* / Fig. 65 *mi.fa 2*	Fig. 66 *si 2*	Fig. 68 *si$^\flat$ 3−*	Fig. 71 *fa$^\#$ 4*	Fig. 76 *do$^\#$ 5* / Fig. 77 *re 5*	Fig. 81 *sol$^\#$ 5+*	Fig. 87 *do$^\#$ 6+* / Fig. 88 *re 6−*
3	Fig. 67 *sol$^\flat$ 3+*	Fig. 69 *si 3* / Fig. 70 *do$^\#$ 4*	Fig. 71 *fa$^\#$ 4*	Fig. 75 *do 5*	Fig. 80 *fa$^\#$ 5*	Fig. 83 *si 5−* / Fig. 84 *do 6−*	Fig. 89 *mi 6*
4	Fig. 72 *sol$^\#$ 4* / Fig. 73 *sol$^\#$ 4+*	Fig. 74 *si$^\flat$ 4−*	Fig. 76 *do$^\#$ 5* / Fig. 77 *re 5*	Fig. 80 *fa$^\#$ 5*	Fig. 82 *si$^\flat$ 5*	*re$^\#$ 6*	Figs. 91 and 92 *sol 6+*
5	Fig. 78 *fa 5−*	Fig. 79 *fa$^\#$ 5−*	Fig. 81 *sol$^\#$ 5+*	Fig. 83 *si 5−* / Fig. 84 *do 6−*	*re$^\#$ 6*	*fa$^\#$ 6+*	*si$^\flat$ 6−*
6	Fig. 85 *do 6−*	Fig. 86 *do 6*	Fig. 87 *do$^\#$ 6+* / Fig. 88 *re 6−*	Fig. 89 *mi 6*	Figs. 91 and 92 *sol 6+*	*si$^\flat$ 6−*	

(considered as a rectangle where the width is equal to the length, and where, as a consequence, the direction of the lines does not make any difference) to those of a rectangle (where the width is less than the length).

I have attributed here the sound *sol 1*, in the mode of vibration of 1|1, because it seems to me to be a product of 2 and 3, in comparison to the other sounds, and because I look at every *do* as some power of 2, according to Par. 5.

The series of sounds that are in agreement with the mode of vibration, where there are nodal lines only in one direction, 2|0, 3|0, 4|0, 5|0, is the same as that of a strip or a band where both ends are free (Par. 75) and is equal to the squares of 3, 5, 7, 9, etc. In the cases where the same number of nodal lines are shown in two different ways, for example, 2|0 and 3|0, the figures where the nodal lines are curved inwards (Figs. 64 and 72) are more in accordance with the exact ratio than those where the lines are curved to the outside (Figs. 65 and 73).

The relative numbers of vibrations for a number of nodal lines going in only one direction, (2|0, 3|0, 4|0, etc.) being equal to $3^2, 5^2, 7^2$, etc., it seems that the sounds of the figures where there are also nodal lines in the other directions will be products of 3, 5, 7, etc. and of another number. The sounds of the series 1|1, 2|1, 3|1, etc. seem to be the products of 3 and the numbers 2, 5, 10, 17, where every second difference is 2; but this progression does not extend to 4|1; the series 2|0, 2|1, 2|2, 2|3, etc. seems to be (by reasoning) the products of 3 and the numbers 3, 5, 9, 15, 23, where every second difference is 2; the sounds of the series 3|0, 3|1, 3|2, 3|3, etc. seem to be the products of 5 and the numbers 5, 6, 9, 13, 18, 24, where, in counting to 3|1, every second difference is 1. The sounds of the figures, where there are the same number of nodal lines in both directions (2|2, 3|3, 4|4, etc.), seem to be between (by reasoning) the squares of 2, 3, 4, etc. except that of the first one, 1|1.

But it is only conjecture; perhaps the true numbers, which these numbers approach, are much more complex. I guarantee only the results of the experiments, explained in the Table (and continuing in this paragraph and in the other Tables). The differences that one will be able to find by well-done experiments are never greater than a semi-tone, more or less. If, therefore, the results of any theory are not conforming to those of the experiments (as, for example, the ratios of the sounds of a square plate that Jacob Bernoulli has given in the *Aeta Acad. Petrop.* 1787); the proposed theory is not the true one. It will be necessary, therefore, to look for another theory that is noted by experiment. And in that case, where one does not succeed, it will be much more advantageous to admit when it surpasses the means furnished by the actual state of science, or the individual faculty who leads the readers astray by a theory that does not conform very well to nature.

109. Several Other Kinds of Vibrations in Which the Plate is Not Free

In all of the types of motion mentioned here, the plate is regarded as being freely vibrating; but there are other motions which are different than these, such as the vibrations of a rod or strip where one or both ends are fixed, different from the same rod or strip where both ends are free.

A complete examination of all these motions would have entailed too much verbosity; this is the reason why I limit myself to mentioning the two I have observed most often. One can only judge them inexactly, and not fit them into the series of sounds of a freely moving square plate.

Figure 97 shows the plate being held tightly between the ends of the thumb and of another finger near n, leaning the corner m on a stationary object, and the bow is applied at the corner p. This manner of motion is the same thing for a square plate, viewed in the diagonal direction as a rhombus, and for the first mode of vibration for a rod where the end is supported (Fig. 22). The sound is graver by a minor seventh, as that of $1|1$ (Fig. 63).

This is also the kind of motion belonging to Fig. 98 that is produced, in effect, in almost the same manner as that produced in Fig. 97, except that the point, where one squeezes the plate, has to be closer to the corner where one applies the bow. The sound is more acute by an octave that the one of $1|1$ (Fig. 63).

110. Different Patterns That Are Formed When the Plate is Not Free

Given what has been said (Par. 96) on the relationships of the figures in general, and on the drawings that are formed by their combinations, a preference can be shown for square plates. If one combines four plates of the same size, on which one has produced the same simple figures; this combined figure will also be produced, more or less, exactly the same on a single larger plate. And in combining several plates (on which are located the same figures), it will form the following drawings:

1. If the lines (parallel or diagonal on the sides) that intersect under a right angle, in Figs. 63, 64, 65, 66a, 67c, 68a, 71a, 72a, 73a, 75, 80a, 81a, 82, etc. are composed of four times the same simple figure, to form a more complex figure, one will have:

Figure 63 four times, while making Fig. 68a
Figure 64 . . . . .72a
Figure 65 . . . . .73a
Figure 68a 80a

2. As in #1, and in each square a small circle, for example, Figs. 70, 76, 77, 88b, 89b, 93, and 94:

Figure 70 four times, while making Fig. 88b
Figure 76 93
Figure 77 94

3. As in #1, but within each square one finds a figure with four curves which encloses a small square with rounded corners. This drawing is reproduced in Figs. 71b, 91, and 92. The last two, when combined four times, also form Fig. 71b in two different ways.
4. As in #1, and within each square there is a figure of the same nature, as in #3, but much more complex, as in Fig. 80b, which when combined four times, gives Fig. 95 or 96.
5. Almost as #3, but the figure contained within the square is another situation, and every point where two lines intersect is nearby a circle (Figs. 74b and 84).
6. The parallel lines on the sides, and the diagonals which intersect at the same point (Figs. 69 and 87b), are also formed when taking Fig. 69 four times.
7. As in #6, and each point where two lines intersect, is nearby a circle, or an almost-square, with rounded corners (Fig. 79a).
8. As in #6, and within each square is found an ellipse, or a line, continuing within the square (Fig. 83), passing by the small axis.
9. As in #6, each point where four lines intersect, it is near a circle, and within each triangle is found a small circle (Fig. 90).

The combinations of some other figures will be able to produce more complex drawings.

111. Symbols for Expressing the Vibrations of Rectangular Plates

The sounds of rectangular plates, where the width is less than the length, have ratios other than those of a square plate. The only exception is the series of vibrations similar to the transverse vibrations of a rod or strip where the ends are free (Par. 71), and where the sounds do not depend on the length. For determination of the sounds of the rectangular plates according to different ratios of the width and the length, I will look at one of these dimensions (the length) as constant, and the other (the width) as variable.

The same plates, being square, having given the sounds mentioned (Par. 108), will give the sounds contained in the following Tables (after one or the other dimension is decreased). I will express the numbers of the nodal lines in the same manner as the preceding, in separating them with a vertical line. The first number expressed is always the transverse line (or parallel to the smaller dimension) and the second number always the longitudinal line (or parallel to the bigger dimension).

For those that are not knowledgeable, it is necessary to note that if the plates give other sounds that those that are contained in the Tables; it is necessary to *transpose* all of them, for the ratios remain the same.

112. Passages of One Figure to Another, When the Sound Is the Same

When the ratio of the dimensions of a rectangular plate is such that two modes of vibrations give the same sound, these different modes of vibration will be able to be represented by an intermediate twisting figure, where one would be able to make small changes in the points where it is touched or rubbed, in one or the other ways of making the vibrations more pronounced, without any change to the sound.

Sometimes, if the sounds of two types of vibrations are slightly different, one of these figures will also be able to pass into the other, and the sound will be a little more raised or lowered, according to whether the figure yields itself more or less to one type of motion or the other.

113–123. Sounds of Rectangular Plates with Different Ratios of Their Dimensions, Regarding the Length as Constant and the Width as Variable

113. When the same square plate, of which the sounds are mentioned (Par. 108), is shortened on a side, in a way that the *width is to the length as 8–9*, it gives the following sounds:

				Number of transversal lines			
		0	1	2	3	4	5	6
Number of longitudinal lines	0			$re^\#$ 2	la 3	la 4−	fa 5+	do 6
	1		la 1	$do^\#$ 3+	$do^\#$ 4	si 4	$fa^\#$. . .sol 5	$do^\#$ 6
	2	sol 2+	re 3+	do 4	$sol^\#$ 4+	$re^\#$ 5+	si^b 5	mi 6
	3	$do^\#$ 4	mi 4	la 4+	$re^\#$ 5+	$sol^\#$. . .la 5	$do^\#$. . .re 6	$fa^\#$ 6
	4	$do^\#$ 5−	re 5	fa 5	la 5+	$do^\#$. . .re 6	fa 6	
	5	la 5	si^b 5	do 6	$re^\#$ 6	$fa^\#$ 6	si^b 6	

The series of sounds of the most simple vibrations, where there are only nodal lines lengthwise, which I express here as $2|0$, $3|0$, $4|0$, etc. is, as on a square plate, and on every rod or narrow strip, according to Par. 71, equal to the squares of 3, 5, 7, 9, etc. But the sounds $re^\# 2$, la 3, la 4−, fa 3+, and do 6 are more acute by nearly a semi-tone, than those of a square plate (Par. 108), although, following the theory, they do not depend on the width.

Nevertheless, the increase of the frequency of these sounds is insignificant. And as to the mode of vibration, it shows two different modes of vibration on one square plate, as $2|0$ and $4|0$. One will be able to look at this small change rather as an average term between the sounds of the two different modes of vibration, when the number of lines is the same.

The sounds of the modes of vibration, where there are no nodal lines lengthwise and where $0|2$, $0|3$, $0|4$, etc. are like the squares of 3, 5, 7, 9, etc., are more acute than the ratio of the squares of the dimensions because they are dependent on the smaller of the two dimensions; or close to 64:81.

In the same mode of vibration $1|1$, the sound, which is always in the inverse ratio of the larger of the surfaces, rose to 9:8. All the other sounds are also more acute than those of a square plate; the difference is more considerable in the mode of motion that presents more lines lengthwise than in the one that presents a larger number widthwise.

114. When the width of the same plate is again smaller, in the manner that it has the *length as 5 to 6*, the plate will give the following sounds:

		Number of transversal lines					
		0	1	2	3	4	5
Number of longitudinal lines	0			$re^\# 2+$	$la\ 3+$	$la\ 4$	$fa\ 5+$
	1		$si^b\ 1$	$re\ 3$	$do^\# 4+$	$si\ 4-$	$sol\ 5-$
	2	$la\ 2+$	$mi\ 3$	$do^\# 4+$	$la\ 4$	$mi\ 5$	$si^b\ 5$
	3	$re^\# 4+$	$fa^\# 4$	$si\ 4-$	$mi\ 5$	$si^b\ 5$	
	4	$re^\# 5$	$mi\ 5$	$sol\ 5$	$si^b\ 5+$		

The series of sounds with the simpler vibrations $2|0$, $3|0$, $4|0$, etc. equal to the squares of 3, 5, 7, 9, etc. is almost the same as that in the preceding paragraph. The one of $0|2$, $0|3$, $0|4$, etc., being in the same ratios, is higher due to reversing the squares of the dimensions, and $1|1$ is higher than the simple inverse of the widths of the surfaces. All the other sounds are also raised more or less. All that being the same in the following cases, I will not repeat it each time.

$4|1$ and $2|3$ can transform themselves into each other through Figs. 157a–e, without any change to the sound, as I have noted in Par. 112. Also $4|2$ and $1|4$, which give the same sound, can pass from one to the other.

115. If *the width is to the length, as 4 is to 5*, the sounds of the same plates will be:

		Number of transversal lines					
		0	1	2	3	4	5
Number of longitudinal lines	0			*re*# 2+	*la 3+*	*la 4*	*fa 5+*
	1		*si 1*	*re*# 3−	*re 4+*	*si 4+*	*sol 5+*
	2	*si 2+*	*fa 3+*	*re*# 4−	*si*b 4−	*fa 5*	*si 5*
	3	*fa 4+*	*sol 4+*	*do 5+*	*fa 5+*	*si*b 5+	*re*# 6
	4	*fa 5*	*fa 5+*	*sol*#. . .*la 5*	*do 6+*	*re*# 6+	*sol*# 6

Here the figures 5|0 and 1|4, which give the same sound, can pass from one to the other, through Figs. 158a–c. The outside lines of 5|0, which are curved inwards twice in Fig. 158a, are also able to curve outwards twice, as in Fig. 159, with a slightly significant raising of the sound. Figure 159 also can transform itself into distortions of 1|4. The same figure also can pass into 3|3, where the sound is the same. Figures 160 and 161 can represent 0|4 and 4|2, and pass into the more pronounced shapes of these two modes of vibration.

116. In the ratio where the *width is to the length as 5 is to 7*, one could presume that the mode of vibration 4|0 and 0|3 would give the same sound and could pass from one to the other because the sounds of the series 2|0, 3|0, 4|0, etc. depend on the length, and those of the series 0|2, 0|3, 0|4, etc. depend on the width, and because the sounds of each series are among them as squares of 3, 5, 7, etc. The different frequencies that belong to the same motions are made up by the difference of the dimensions themselves and the motions of their ratios, so that both methods of vibration, 4|0 and 0|3, have to be equal to $5^2 \times 7^2$; that which is noted by the experiment. These two types of vibrations are represented most easily by the intermediate Fig. 163b, which can be transformed, by small changes in the place of touching, into Fig. 163a or into Fig. 163c, and sometimes also in three straight lines, according to the length, or four according to the width, without any change to the sound.

Given that the ratio of the dimensions 5–7 is also very close to that of 1 to the square root of 2, one could suppose that the sounds of the series 0|2, 0|3, 0|4, etc. would be more acute by an octave than those of the series 2|0, 3|0, 4|0, etc., which also conforms to experience. I used plates of which the ratio of the diameters is between 5:7 and 1: $\sqrt{2}$, which differ only by $\frac{1.4142....}{1.4}$, which is not sensible to the ears or the eyes. Here are the ratios of the sounds of these plates:

		Number of transversal lines					
		0	1	2	3	4	5
Number of longitudinal lines	0			re# 2+	la 3+	la 4	fa 5+
	1		do# 2	mi 3−	re# 4	do 5	sol 5−
	2	re 3+	sol# 3	mi 4+	do 5	fa# 5	do 6−
	3	la 4+	si 4−	re# 5	sol# 5−	do 6−	mi 6
	4	la 5	la 5+	si 5−	re# 6	fa 6+	

The mode of vibration 1|2 is shown as standard, as in Fig. 164b, which, if the ratio of the dimensions is altered slightly, can change into figures, or show the lines as separate, curved, or straight. If the width is slightly diminished, it can pass into 3|0, as I have shown in Figs. 164a, c. Here 3|0 is shown as standard, as in Fig. 165a, which, if the width is slightly diminished, can pass into 1|2 with straight lines, as in Fig. 165b. When the ratio of the dimensions is slightly changed, 3|3, represented by Fig. 166, can pass into 5|1, and 5|1, represented by Fig. 167, can pass into 3|3.

117. If *the width and the length are as 2 to 3*, or rather, if the width is again diminished slightly, 2|0, which represents the first mode of transverse vibrations, and 1|1, which represents the first mode of rotating vibrations, give the same sound (Par. 98), and can pass from one to the other by the distortion of the nodal lines.

4|1 can pass into 0|3, as in Figs. 168a–c, and Figs. 169a–c.

118. If *the ratio of the width to the length is as 3 is to 5*, one could presume, for the same reasons as for 4|0 and 3|0 in Par. 116, that 3|0 and 0|2, both equal at $3^2 \times 5^2$, should give the same sound, and be represented by the same intermediate figure; which agrees with the experiment. These two types of motion are most easily represented by Fig. 170b, that can be transformed, by small changes in the place where it is touched, as in Figs. 170a or c, or that can pass into the shapes again more pronounced of 0|2 or 3|0. The ratios of the sounds of a similar plate were:

		Number of transversal lines					
		0	1	2	3	4	5
Number of longitudinal lines	0			re# 2+	la 3+	la 4	fa 5+
	1		fa 2	fa# 3	mi 4+	do 5+	sol 5+
	2	la 3+	do# 4+	sol# 4	re# 5	sol# 5	do# 6
	3	re# 5	mi 5	sol 5	si 5	re# 6	
	4	re# 6−	re# 6	mi 6+	sol 6	la 6	

119. If the *ratio of the two dimensions is as 4 to 7*, or rather, if the ratio is again slightly altered, 4|0 and 2|2, which give the same sound, can pass from one to the other through Figs. 171a–c, or through Figs. 172a–c; similarly 5|0 and 1|3 can pass from one to the other in the middle of a transverse line, as they do in Figs. 163a–c.

120. The *ratio of the two dimensions being as 1/2 to 1*, the theory and the experiment shows that the sounds 0|2, 0|3, 0|4, etc., which are dependent on the width, are more acute by two octaves than those of 2|0, 3|0, 4|0, etc., which are dependent on the length. The plates produced the ratios of the following sounds:

		Number of transversal lines					
		0	1	2	3	4	5
Number of longitudinal lines	0			$re^\#$ 2+	la 3+	la 4	fa 5+
	1		sol 2+	la 3+	$fa^\#$ 4	re 5	$sol^\#$ 5+
	2	$re^\#$ 4	$fa^\#$ 4	do 5	$fa^\#$ 5	si 5	$re^\#$ 6
	3	la 5	si^b 5	do 6	$re^\#$ 6	$fa^\#$ 6	si^b 6

2|1 and 3|0 pass from one to the other through Figs. 173a–c; if the width is slightly greater than $\frac{1}{2}$, 5|1 and 1|3 can be transformed from one to the other through Figs. 174a–c and give the same sound.

121. The *width being to the length as 3 to 7*, following the theory (Par. 116), and the experiment, 4|0 and 0|2, equal to $3^2 \times 7^2$, produce the same sound, and can pass from one to the other in two different ways, through Figs. 175a–c and through Figs. 176a–c. The sound of Fig. 176 (Par. 106) is slightly more acute than that of Fig. 175.

122. If the *ratios of the diameters are as 1/3 to 1*, the sounds of 5|0 and 0|2, equal to $3^2 \times 9^2$, are the same, conforming to the theory (Par. 106). The two types of vibrations are represented by Fig. 177b, which can pass through Figs. 177a and 177c, and sometimes has two straight lines according to the length, or four according to the width. The ratios of the sounds were:

		Number of transversal lines					
		0	1	2	3	4	5
Number of longitudinal lines	0			$re^\#$ 2+	la 3+	la 4	fa 5+
	1		re 3+	$re^\#$ 4+	do 5−	$fa^\#$5	do 6−
	2	fa 5+	sol 5	si 5	re 6+	$fa^\#$ 6	si^b 6−

123. The *width being $\frac{1}{4}$ of the length*, the ratios of the sounds were:

		Number of transversal lines					
		0	1	2	3	4	5
Number of longitudinal lines	0			$re^{\#}$ 2+	la 3	la 4−	fa 5+
	1		sol 3+	$sol^{\#}$ 4+	mi 5−	si^{b} 5	$re^{\#}$ 6
	2	$re^{\#}$ 6+	mi 6	$fa^{\#}$ 6	la 6		

One sees that the sound of 0|2, conforming to the theory, is more acute by four octaves than that of 2|0. Because of the greater decrease, it was very difficult to produce types of vibrations that offered more than one nodal line according to the length.

The *width being again decreased to be $\frac{1}{6}$ of the length*, the ratios of the sounds were:

		Number of transversal lines					
		0	1	2	3	4	5
Number of longitudinal lines	0			$re^{\#}$ 2+	la 3+	la 4	fa 5+
	1		re 4	re 5+	la 5+	$re^{\#}$ 6	sol 6+

If the *width is only $\frac{1}{8}$ of the length*, the ratios of the sounds were:

		Number of transversal lines					
		0	1	2	3	4	5
Number of longitudinal lines	0			$re^{\#}$ 2+	la 3+	la 4	fa 5+
	1		sol 4	sol 5	re 6	sol 6	si 6

When the width was again further diminished, the series of the sounds of the transverse vibrations, 2|0, 3|0, 4|0, stayed the same, and those with rotating vibrations, 1|1, 2|1, 3|1, etc., approached the natural series of numbers 1, 2, 3, 4, etc., and their absolute frequency increased in the same ratio with which the width was diminished.

124. Summary of the Research on the Vibration of Rectangular Plates

In comparing all of the ratios of the sounds of rectangular plates, where one dimension is constant and the other variable, one will see:

1. That the modes of vibration 2|0, 3|0, 4|0, etc. similar to the transverse vibrations of a rod or a blade where the ends are free (Par. 78), have always kept their ratios as the squares of 3, 5, 7, 9, etc., and their absolute pitch, because they are dependent only on the length. The change of nearly a semi-tone, that is noticed in the passage of a square plate to a rectangular plate of unequal dimensions, is more apparent than real. If one puts the average terms between the sounds of the vibrations on a square plate where the number of the nodal lines are the same, the lines are curved inwards or outwards.

2. That the modes of vibration where the nodal lines are only according to the length, 0|2, 0|3, 0|4, etc., as much as the diminished width will allow them to be produced, have also kept among them the ratios of the squares of 3, 5, 7, 9, etc.; but the absolute pitch is augmented because of the inverse of the squares of the width, conforming to the theory. These vibrations have the same ratio to the width that 2|0, 3|0, 4|0, etc. have to the length.

3. For the mode of vibration where one longitudinal line is intersected by the transversals, 1|1, 2|1, 3|1, etc., the sounds of a square plate are nearly in the ratios of the numbers 6, 15, 30, etc. But if one of the dimensions is slightly diminished, the sounds come closer and closer to the mode in which they finally pass into the natural series of numbers 1, 2, 3, 4, etc. This agrees with the torsional vibrations of a rod or blade, in which the motions are the same here as in the modes of vibration 1|1, 2|1, 3|1, etc. The sound of 1|1 is in inverse ratio to the surfaces; when the width is close to $\frac{1}{3}$ of the length, it is equal to that of 2|0.

4. That in all of the other types of motion, where several nodal lines, in one of the directions, are intersected by the lines of the other direction, the sounds are more acute when the width is diminished, and that the difference is significant if there are several longitudinal lines.

5. That in all cases where two different types of motion give the same or nearly the same sound, the figures of the nodal lines are able to pass from one to the other through the intermediate figures.

D. Vibrations of a Round Plate

125. Nodal Lines in Diametral and Circular Directions, and Symbols for Expressing Them

In all the possible kinds of vibration of a round (free) plate, the *nodal lines* are either *diametral* or *circular*, which can be either regular, or distorted in different ways, without alteration of the ratio of the sound which is suitable for the same type of vibrations. I will express the number of nodal lines in almost the same manner as for rectangular plates, by using a vertical line to separate two numbers. Before the line is the number of nodal lines in the diametral directions and after the line (written in Roman numerals) is the number of nodal lines parallel at the periphery. Thus, for example, 2|0 will express the kind of vibrations in which there are only two diametral lines; 0|I the case where there is only a single circular line; 4|III that in which there are four diametral and three circular lines, etc.

126. Vibrations in Which There Are Only Diametral Lines

When there are only two nodal lines in the diametral directions, these lines can be straight and intersect at the center of the plate. The figure will then be seen in the shape of a star with 4, 6, 8, 10, or more rays; but when these lines intersect in a different way, the number of lines, counting from one end to the other, and the ratio of the sound, remain the same.

2|0 (Fig. 97), where two diametral lines intersect at the center is, of all possible figures, the one that gives the deepest sound. If the plate is small, this figure, as the simplest, will appear more easily when the plate is clamped at its center and the bow is applied at any point; one will also produce it on any coin, if it is not too small. If the plate is larger, one must, in order to exclude other movements, clamp the plate slightly away from the center, or clamp it at the center and touch at the same time another point at which one wishes that a nodal line pass, at 45° from the nodal line determined by the touching. I repeat here the remark that, in order to produce simple figures, which give the deepest sounds, we always use a greater pressure and a softer motion of the bow than that used to produce complex figures that give higher sounds.

The second mode of vibration, 3|0 (Fig. 100), where three diametral lines, which intersect at the center, appear in the shape of a star, gives a more acute sound (by one ninth) than the first. It is produced by touching the plate in the same manner as for the first type of motion, and by applying the bow at a distance of about 90° from the nodal line, determined by touching.

In order to produce other modes of vibration, we can clamp the plate farther away from the center, where there are more nodal lines, because the part in the center of the plate is more stationary. The bow should always be applied in the center of the vibrating part. When one has gotten used to it, one is more likely to discover the most suitable places by a correct look and by trial and error, than by measurement.

4|0 is shown in the shape of a star of eight rays (Fig. 101a) or disfigured, as Fig. 101b. The sound is more acute by a minor seventh than that of 3|0, and by two octaves than that of 2|0.

5|0 appears either as Fig. 102a, or more often distorted as in Fig. 102b. The pitch of the sound increases by almost a minor sixth.

Of all the other figures, 6|0, 7|0, 8|0, etc., that can appear in the shape of a star, or distorted, 8|0 is the most susceptible to regular distortions, of which those that I have seen most often are represented in Figs. 103a, b.

The frequencies of the sounds of these modes of vibration are nearly in the ratios of the squares of the numbers of nodal lines, but the measured ratios seem to be slightly less than these exact ratios.[2]

127. Vibrations That Present a Circular Line

A circular line can be alone or can be intersected by 1, 2, 3, or more diametral lines.

0|I (Fig. 104) gives a sound that is more acute by a superfluous fifth, $4^2:5^2$ than 2|0 (Fig. 99). It is necessary to clamp a point of the circular line between the fingertips, and to apply the bow near the point of touching, employing more pressure and less speed than for the other figures. The motion is the simplest of all those to which a round plate is susceptible because the vibrations of each diameter resemble those of a rod whose ends are free (Par. 71) and which bends to the curvature represented in Fig. 24. The sound of this figure, and of those where there are circular lines, is more sonorous and of a different timbre than that of the figures where there is only one diametral line.

1|I (Fig. 105) is the easiest to produce of all possible figures, if the plate is clamped close to the edge and the bow is applied at a distance of about one quarter of the periphery. The diametral line passes through the touching point. Therefore, by varying this point, we can change the position of this line at will. The sound is more acute by a ninth than that of 0|I.

---

[2] Chladni correctly observed that the frequencies for modes 3|0, 4|0, and 5|0 are roughly 9, 16, and 25 times the frequency of mode 2|0. The second part of Chladni's observation is also correct. More accurate values for these ratios are 8.4, 15.0, and 23.9.—*JPC*

2 | I (Fig. 106) will appear if you shake the plate as to produce 1|I, and if the bow is applied at a point as little as 45° from the touching point. The sound is more acute by almost a seventh minor than 1|I and by two octaves than 0|I.

In order to produce 3|I (Fig. 107), 4|I (Fig. 108), and 5|I, we clamp a point near the edge through which one wishes a diametral line to pass, and apply the bow as closely as possible to the touching point, where the number of diametral lines that can be produced is greater. To make the figure more pronounced, we can also touch another diametral line below at the same time with a fingertip, where the circular line is as far removed as possible, so that the number of diametral lines will be greater. The position of the lines can also be fixed by lightly touching the edge at a point where a diametral line terminates, where it is susceptible to yielding slightly at this obstacle.

128. Vibrations in Which There Are Two or More Circular Lines

Two or more circular lines can also be either alone or intersected by diametral lines. The circular lines can appear as concentric circles; they can also take on a certain number of flexures and resemble epicycloids. Flexures of two circular lines ordinarily approach and recede from one another; they are less pronounced in inner circular figures than in the outer ones. The inner circle ordinarily takes an elliptical shape. When the diametral figures are distorted, they often resemble hyperbolas.

0|II, if we wish that the circular lines be concentric (Fig. 109a), is a little difficult to produce, as are all the figures where the lines do not intersect at any point, because it is necessary to touch the single lines very exactly (and not too far beyond) in order that the vibrations of the neighboring segments not be too obstructed. However, one can produce this figure on each plate that is not too small. One must clamp a point on an outer circle between the tips of the thumb and another finger, at the same time touching the inner circle with the tip of another finger, and applying the bow sufficiently strongly and slowly, close to the clamped point, in such a manner that the points of touching and stroking are in the same semi-diameter. This mode of vibration is the simplest after 0|I, because each diameter makes its motions as a free rod (Par. 71), in the mode of vibrating where there are four vibration nodes. The sound is more acute by two octaves than that of 0|I; the ratios of the sounds therefore differ a great deal from those of a rod when it vibrates in the same manner. One can produce much more easily, even on smaller plates, the distortion of 0|II represented in Fig. 109b, where the outer circle is bent five times and the inner circle is oval. We must support the plate with two fingertips against some obstacle that is not too hard, so that these three supported points correspond to points where the flexures of the outer circle touch the edge; one then applies the bow to the edge at a point where the line is bent inwards. The sound of this distortion is ordinarily slightly more acute than that of the regular form; the difference can be almost a semi-tone. When the circles are concentric, one can always note a tendency of the outer circle to bend five times and of the inner circle to take an elliptical oval shape. On a brass plate that was not very exact, the outer circle was always curved six times and the inner was elliptical.

1|II, with concentric circles (Fig. 110a), can be produced by operating in almost the same way as in the production of 1|I, but it is necessary to clamp the plate slightly closer to the edge, and to touch (at the same time) one or two points of the outer or inner circle. The same type of vibrations can also be slightly transformed in such a way that the outer circle is bent six times, and the inner circle becomes elliptical (Fig. 110b), if one operates it almost in the same way as to produce the distortion of 0|II, of which I have spoken.

2|II can appear regular, as in Fig. 111a, if we operate it almost as in the production of 1|II, but in applying the bow to a point that is closer to the clamping; if we attach the plate in three points, the outer circle is bent six times, and the shape is changed into that of Fig. 111b or c.

3|II appears occasionally regular, but ordinarily it is transformed into Fig. 112a or b; 4|II regular, as in Fig. 113a or b; 5|II is represented in Fig. 114 such that the diametral lines are intersected regularly in the middle; but in this mode of vibration, and in the others, they can also take on other shapes.

0|III appears very rarely as Fig. 115; 1|III appears as Figs. 116a, b; 2|III as Figs. 117a, b; 3|III as Figs. 118a, b; 4|III as Figs. 119a, b, etc. The circular lines were sometimes more concentric.

In Figs. 120 and 121a, I have represented the distortions of 0|IV and 1|IV, where two considerable points are without motion, such that the sand remains at rest but does not accumulate as on the nodal lines; 0|V, 1|V, 0|VI, 1|VI, 0|VII, etc. are susceptible to similar more complex distortions; 1|IV also appears as Fig. 121b or c.

When there are circular lines, the diametral lines ordinarily take on the same shapes as in the figures represented in the drawings.

One cannot always produce at will these complex figures. If, however, one makes use of plates that are large enough, regular, and of slight thickness, and if one varies the manner of operating by clamping or touching the plate, by supporting it, or by placing it on small trestles of cork, and by applying the bow at different points, etc. one will often succeed in producing sufficiently exact, rather complex figures. But it should be observed that the bow does not move to the right or to the left, and that the degree of pressure and the speed of motion of the bow remain the same in order that the sound does not change and that the figure appears sufficiently pronounced.

129. The Bending of Circular Lines

In order to compare the number of flexions in the circular lines, I have put together the following table:

		Number of diametral lines								
		0	1	2	3	4	5	6	7	8
Number of circular lines	II	5	6	6	7	7	7	8	8	8
	III	8	9	9	10	10	11	11	11	11
	IV	12	12	13	13	13	14	14		
	V	15	15	15?	16					
	VI	18?	18	19	19					
	VII	21								

It is seen that, ordinarily, the number of flexures of the circular lines does not have such a ratio to the number of diametral lines that one can be divided by the other or that there exists a common divisor. Consequently, there are no points of perfect symmetry in these figures. It is this that makes the figures difficult to draw; if one wished to give to the diametral lines the same position on both sides, the flexures would become too unequal. But if one would wish to give to all the flexures the same size, the position of the diametral lines would have too little symmetry. In both of these cases, the figure would not be as nature gives it, where each vibrating part has the size compatible for it to be in equilibrium with all the others. Consequently, these figures seem to be more symmetric than they really are.

> On the brass plate already cited, where the outer circular line in 0|II has six flexions, there are nine in 0|III and ten in 1|III.

130. Ratio of Sounds of a Round Plate

The sounds of a round plate, if one assigns *do 1* to the mode of vibration that gives the deepest sound, 2|0 (Fig. 99), will be the following:

Number of circular lines	Number of diametral lines								
	0	1	2	3	4	5	6	7	8
0			Fig. 99 *do 1*	Fig. 100 *re 2*	Fig. 101 *do 3*	Fig. 102 *sol. . .sol*$^{\#}$ *3*	*do*$^{\#}$ *4*	*fa*$^{\#}$ *4*	Fig. 103 *si*$^{\flat}$ *5*
I	Fig 104 *sol*$^{\#}$ *1*	Fig. 105 *si*$^{\flat}$ *2*	Fig. 106 *sol 3*	Fig. 107 *re. . .re*$^{\#}$ *4*	Fig. 108 *sol*$^{\#}$ *4*	*do*$^{\#}$ *5*	*mi. . .fa 5*	*sol 5*	
II	Fig. 109 *sol*$^{\#}$ *3+*	Fig. 110 *mi 4+*	Fig. 111 *si*$^{\flat}$ *4*	Fig. 112 *re*$^{\#}$ *5*	Fig. 113 *sol 5*	Fig. 114 *si*$^{\flat}$*. . .si 5*	*do*$^{\#}$ *6*	*re*$^{\#}$ *6*	
III	Fig. 115 *si*$^{\flat}$*. . .si 4*	Fig. 116 *mi 5+*	Fig. 117 *sol*$^{\#}$*. . .la 5*	Fig. 118 *do 6*	Fig. 119 *re*$^{\#}$ *6*	*fa*$^{\#}$ *6*	*sol*$^{\#}$ *6+*	*si*$^{\flat}$ *6*	
IV	Fig. 120 *la 5*	Fig. 121 *do*$^{\#}$ *6*	*fa 6−*	*sol. . .sol*$^{\#}$ *6*	*si*$^{\flat}$ *6*	*si 6. . .do 7*	*do*$^{\#}$ *7*		
V	*fa 6*	*sol*$^{\#}$ *6*	*si 6*	*do*$^{\#}$ *7*					
VI	*si 6*	*re 7*	*mi 7*	*fa 7+*					
VII	*mi 7*								

The ratios of these sounds correspond approximately to the squares of the following numbers:

Number of circular lines	Number of diametral lines								
	0	1	2	3	4	5	6	7	8
0			(2)	(3)	(4)	(5)	(6)	(7)	(8)
IV	2	3	4−	5−	5−	7−−	8−−	9−−	
II	4+	5+	6	7−	8−	9−−	10−−	11−−	
III	6+	7+	8−	9	10−	11−	12−	13−−	
IV	8++	9+	10+	11+	12	13−	14−		
V	10++	11++	12+	13+					
VII	12++	13++	14+	15+					
VII	14++								

By adding +, I indicate that a sound is slightly more acute, and, by adding −, that it is slightly graver. And when the number is the same but the sign is added twice, it is even sharper or flatter than if the sign is added only once. When the number of the diametral lines is the same, the number of circular lines increases and each interval is slightly larger. When the number of circular lines is the same, the number of diametral lines increases, and each interval is slightly smaller than the squares of these numbers. The series of sounds where there are only diametral lines, $2|0$, $3|0$, $4|0$, etc. corresponds to the squares of 2, 3, 4, etc.; but one must regard this as separated from all the other series. For this reason, I have expressed them by the squares of (2), (3), etc. In the other modes of vibration, if one wishes to neglect the alteration of the intervals, due to the preponderance of circular or diametral lines, one could then count a circular line for two diametrals. By expressing by D the number of diametral lines, and by C the number of circular lines, the relative frequency of the vibrations will be $(D + 2C)^2$.[3]

[3] Lord Rayleigh refers to this formula as "Chladni's Law." In fact, we have found that it holds up quite well for other structures with circular symmetry such as church bells.—*TDR*

It should be noted that one cannot easily produce all these sounds on the same plate. Small plates serve better for producing the modes of vibrations that are simplest and large plates for producing those that are more complex. I make use of plates whose diameter is from five decimeters down to a decimeter, and I have transposed the different sounds in order to reduce them to the same size.

131. Some Other Types of Vibrations in Which the Plate Is Not Free

One can also produce several other modes of vibration which do not belong to the series of vibrations of a round free plate. These (like those of a square plate of which I spoke in Par. 109) differ just as the vibrations of a rod with one end attached (Par. 70) differ from those where the ends are free (Par. 71). Several forms of nodal lines, which correspond to this case, are represented in Figs. 122–126. The point where one must attach the plate is marked by points on the edge, which represent the sand that accumulated near this point, as on the nodal lines. The point where one holds the plate is marked by n, and that where the bow must be applied by p. Figure 122 is, for a round plate, the same thing as the first mode of Fig. 22; the sound is slightly graver by a major sixth than that of Fig. 99. The sound of Fig. 123 is more acute than that of Fig. 122 by almost an octave and a tone; that of Fig. 124, by almost two octaves and a major third; that of Fig. 126 (which can pass into Fig. 105) by two octaves and a minor sixth.

E. Vibrations of Elliptical Plates

132. General Remarks

The vibrations of elliptical plates will be treated here in the same fashion as those of rectangular plates, assuming one of the axes as constant and the other as variable, beginning with those of a round plate (as an ellipse whose axes are equal), and passing through ellipses that are more and more elongated to the vibrations of a rod or thin sheet.

If the two axes of an elliptical plate differ very little, the vibrations resemble a great deal of those of a round plate; but if the difference of the two axes is quite considerable, they resemble those of a rectangular plate. The possible figures of the nodal lines consist of:

1. Transverse lines; ordinarily, they are bent inwards, the outer ones more than the inner ones, and resemble opposed hyperbolas
2. A longitudinal line in the major axis
3. Elliptical lines that are more elongated than the shape of the same plate

To rank in the most suitable manner all the modes of vibration of an elliptical plate, one can regard each elliptical line as two longitudinal lines curved outward because of the shape of the plate. I will express, for greater precision, the numbers of the nodal lines in the same manner as for rectangular plates, separating them by a small vertical line; the first number will express the transverse lines, and the second the longitudinal, counting an ellipse as two lines. We will therefore have the following series:

1. Modes of vibration where there are only transverse lines, $2|0, 3|0, 4|0$, etc. (Figs. 179–182).
2. A longitudinal line in the major axis, intersected by one, two, three, or more transverse lines, $1|1, 2|1, 3|1$, etc. (Figs. 183–187).
3. One elliptical line which can be regarded as two longitudinal lines, or singles, or intersected by the transverse lines, $0|2, 1|2, 2|2, 3|2$, etc. (Figs. 188–193).
4. An elliptical line and a longitudinal line in the major axis, which is equivalent to three longitudinal lines or singles or intersected by transverse lines, $0|3, 1|3, 2|3, 3|3$, etc. (Figs. 194–199).
5. Two elliptical lines, which can be regarded as four longitudinal lines, or singles (Fig. 200) or intersected by transverse lines, $0|4, 1|4, 2|4, 3|4$, etc.

Likewise, one can show two elliptical lines and a line in the major axis, which is equivalent to five longitudinal lines or three elliptical lines, etc., and in all these cases they can be singles or intersected by transverse lines which appear under the same shapes as if they were singles.

We must remark here the simplest manner of tracing the ellipses of all the ratios of one axis to the other, for those who do not know it. Draw two lines (Fig. 178) that intersect at a right angle; the first, pq, equal to the major axis, and the second, cd, equal to the minor axis. Or, use a compass to find the center of the long axis. Put one end of the compass at the end of the minor axis and mark the points m and n (the foci of the ellipse) where a circle traced with the other end of the compass intersects the major axis. Attach a string to these points, taut, but long enough that a pencil that rests against the string can also touch the ends of the axes; the curve that can be traced with the pencil in this way produces the ellipse.

133. Manner of Producing Different Kinds of Vibrations

To produce the first series of sounds, where there are only transverse lines, $2|0$, $3|0$, $4|0$, etc. (Figs. 179–182), we must grasp the middle of the outer line with the fingertips and apply the bow to the end of the major axis. The sound of these motions is ordinarily harsh and without resonance, because one cannot grasp the plate, on a line that is no more than a point in width, without generating the vibrations of the neighboring segments.

If one wants to produce vibrations where a longitudinal line is intersected by some transversals, $1|1$, $2|1$, $3|1$, etc. (Figs. 183–187), one clamps a point where two nodal lines intersect. The pressing is done for the first figure (Fig. 183) in the middle, and, for the others, in those points on the major axis that are further and further removed from the center; the bow must be applied between the ends of the two lines. This series of figures resembles a great deal the figures of a rectangular plate when the ellipse is strongly elongated.

The mode of vibration in which there is only a single elliptical line, $0|2$, could be produced if the point where this line is clamped, and that where one applies the bow, are approximately on the minor axis. If the shape of the plate is a strongly elongated ellipse, there is ordinarily at each end, where the two longitudinal lines come together, a very considerable point, which remains fixed while the vibrations are transmitted only to the transverse. In this case, it will be suitable to clamp the plate at a point which is not too far from an end, to touch (at the same time) a point of the outer nodal line, more distant from the end, with a fingertip, and to apply the bow close to this touched point, not far from the middle of the long side. To produce the modes of vibrations in which the longitudinal lines are intersected by the transversals, it is necessary to clamp one of the outer points, where two lines intersect, between the ends of the two lines. The more transverse lines that one wishes to produce, the closer the clamped point must be to the edge of the plate. The more longitudinal lines that one wishes to produce, the closer the clamped point must be to where the outer longitudinal lines can be touched with a fingertip, close to the edge. A quick fair look and a little experience make the rest understood.

134. Passages of the Figures of a Round Plate to Those of an Elliptical Plate

It would be useful to show the passage of the figures of the nodal lines on a round plate, where the axes are equal, to their transformations on elliptical plates where the axes are unequal.

The first series of vibrations, where there are only transverse lines, and the second, where one longitudinal line is intersected by some transversals, are the same for an elliptical plate that vibrates, as for a round plate where there are only diametral lines, with one exception. In a round plate, the position of these lines is indifferent, because each diameter is equal to all the others; but in an elliptical plate, the number of lines being the same, the figures and the sound differ a great deal, according to whether all three lines are transversals or whether one of these lines is found on the major axis. When the axes differ very little, the difference of the figures and of the sounds in these two cases will not be detectable. The figures will not be sufficiently pronounced for us to determine the position of the line, which can be changed by small changes in the point of touching without a sensible alteration of the sound. The figures where there are several nodal lines then appear slightly more like Fig. 202, in such a way that the lines are more pronounced than near the edge, and a large part in the middle remains stationary. When the size of the ellipse is reduced more and more, the figures of the simpler modes of vibration begin to relate themselves to one axis or another and to be distinguished by different sounds. When the difference of the axes increases still further, the more complex figures also begin to relate more distinctly to one or the other dimension.

The types of vibrations of a round plate, where there are only circular lines, can only appear on an elliptical plate in one way. But if the circular lines are intersected by diametral lines, each of these types of vibrations can appear on an elliptical plate in two different ways, according to which of these lines is found on the major axes or whether all are transverse.

The transformations of the figures of a round plate would produce the following figures on the elliptical plates:

2|0, produces	Fig. 99	or 2|0, Fig. 179	or 1|1, Fig. 183
3|0, ...	Fig. 100	or 3|0, Fig. 180	or 2|1, Fig. 184
4|0, ...	Fig. 101a	or 4|0, Fig. 181	or 3|1, Fig. 185
			etc.
0|I, ...	Fig. 104	0|2, Fig. 188	
1|I, ...	Fig. 105	or 1|2, Fig. 189	or 0|3, Fig. 194
2|I, ...	Fig. 106	or 2|2, Fig. 190	or 1|3, Fig. 195
3|I, ...	Fig. 107	or 3|2, Fig. 191	or 2|3, Fig. 196
			etc.
0|II	Fig. 109	0|4, Fig. 200	
1|II	Fig. 110a	1|4, or 0|5, etc.	

135. Some Particularly Remarkable Ratios of Axes

The most remarkable ratios of one axis to another are those of 5:3, 8:3, 11:3, 14:3, 17:3, etc. In this series, the number before the vertical line can be expressed as $3n-1$, where n is any integer beginning with 2. In the plates of these ratios of axes, the sounds of the whole set of vibrations (except those where there are only transverse lines, 2\|0, 3\|0, 4\|0, etc.), join together in only one series. If T expresses the number of transverse lines and L the number of longitudinal lines, all the modes of vibration, where the number $T + nL$ is the same, give the same sound.

136–146. Vibrations of Elliptical Plates in Different Ratios of Axes, Regarding One Axis as a Constant and the Other as a Variable

136. When *the ratio of the axes is as 9 to 8*, one can produce the following sounds on the same plate which, being round, would have given the sounds in the ratios of Par. 130:

		Number of transversal lines						
		0	1	2	3	4	5	6
Number of longitudinal lines	0			re 1	fa 2	re 3+	si^b 3	mi 4−
	1		re 1+	fa 2	re 3+	si^b 3	mi 4−	
	2	si 1−	do 3−	si^b 3	fa 4−	si^b 4		
	3	re 3	si^b 3	fa 4−	si^b 4	re$^\#$ 5		
	4	do$^\#$ 4	sol 4−	do 5+				
	5	sol$^\#$ 4+	do$^\#$ 5+	fa$^\#$ 5				

The first series, where there are only transverse lines, is still not separated from the second, where there is one longitudinal line, since the position of the lines is still indifferent and undetermined. In the first mode of vibration 0|2 or 1|1, one can detect a slight rise in the sound if one of the lines is found on the major axis.

Several figures, such as 4|1, 2|2, and 1|3, which give the same or nearly the same sound, can pass from one to the other by intermediate distortions.

137. When the size is diminished a little further, so that *the ratio of the axes is as 5 to 4*, the sounds that can be produced on the same plate will be:

		Number of transversal lines						
		0	1	2	3	4	5	6
Number of longitudinal lines	0			re 1+	fa$^\#$ 2−	mi 3−	si 3−	mi 4+
	1		mi 1	fa$^\#$ 2	mi 3−	si 3−	mi 4+	
	2	do$^\#$ 2+	do$^\#$ 3	si^b 3+	fa$^\#$ 4−	si 4	mi 5	
	3	fa 3	do$^\#$ 4	sol 4	do 5+	fa 5		
	4	mi 4	la 4	re 5				
	5	do 5						

138. The sounds of the same plate, when *one axis is to the other is as 4 is to 3* will be:

		Number of transversal lines						
		0	1	2	3	4	5	6
Number of longitudinal lines	0			*re*# *1−*	*fa*# *2*	*mi 3*	*si 3+*	*fa 4*
	1		*fa 1−*	*fa*# *2−*	*mi 3*	*si 3+*	*fa 4*	
	2	*re*# *2+*	*re 3+*	*si 3−*	*fa*# *4−*	*si 4+*	*mi 5+*	
	3	*sol 3−*	*re 4+*	*sol*# *4−*	*do*# *5−*	*fa 5*		
	4	*fa*# *4−*	*si 4−*	*re*# *5+*				
	5	*re 5*						

The first two series, 2|0, 3|0, 4|0, etc. and 1|1, 2|1, 3|1, etc., are still not completely separated; the difference is almost as perceptible as the first two sounds.

139. *The ratio of the two axes being as 3 to 2*, the sounds of the same plate will be:

		Number of transversal lines						
		0	1	2	3	4	5	6
Number of longitudinal lines	0			*re*# *1−*	*fa*# *2+*	*fa 3−*	*do 4*	*fa*# *4+*
	1		*fa*# *1+*	*sol*# *2*	*fa*# *3−*	*do*# *4−*	*sol 4*	
	2	*fa*# *2+*	*fa 3−*	*do 4*	*fa*# *4+*	*do 5*	*fa 5*	
	3	*si*♭ *3+*	*fa 4−*	*si*♭ *4+*	*re*# *5*	*sol 5*		
	4	*la 4*	*do 5+*					
	5	*fa 5+*						

At present, the figures of the first two series are more pronounced and the sounds are different.

In this ratio of the two axes, several figures that give the same sound can pass from one to the other; as, for example, 3|0 and 0|2, which can be represented by Fig. 203, which, by small changes of the points of touching and of rubbing, can be transformed into three transverse lines or into two longitudinal lines to which an elliptical line is equal, without changing the sound.

It should be noted that in this ratio of axes the figures, where there are two longitudinal or elliptical lines, 0|2, 1|2, 2|2, 3|2, etc., give the same sound as the first series where there are only transverse lines, counting from the second mode of vibration, 3|0, and that the sound of 0|2 is more acute by an octave than that of 1|1.

140. *The ratio of the axes as 5 to 3* is the first degree where the sounds of the figures are reunited (Par. 135) in order to form a single series, except in the case where there are only transverse lines. The same plate, which would have given the sound mentioned, will give the following sounds in this ratio of the axes:

		Number of transversal lines								
		0	1	2	3	4	5	6	7	8	9
Number of longitudinal lines	0			*re*$^\#$ *1*	*sol 2*	*fa 3*	*do*$^\#$ *4*	*sol 4*			
	1		*sol*$^\#$ *1+*	*la 2+*	*sol 3−*	*re 4−*	*sol*$^\#$ *4−*	*do*$^\#$ *5−*	*fa 5+*	*la 5−*	*do 6*
	2	*la 2+*	*sol 3−*	*re 4−*	*sol# 4−*	*do*$^\#$ *5−*	*fa 5+*	*la 5−*	*do 6*		
	3	*re 4−*	*sol*$^\#$ *4−*	*do*$^\#$ *5*	*fa 5+*	*la 5−*	*do 6*				
	4	*do*$^\#$ *5−*	*fa 5+*	*la 5−*	*do 6*						
	5	*la 5−*	*do 6*								

By using T to express the number of transverse lines, and L the number of longitudinal lines, and by supposing $n = 2$, all the modes of vibration in which the quantity $T + 2L$ is the same give the same sound. In the following Table, I will arrange vertically the modes of vibration whose sound is the same.

	1|1	2|1	3|1	4|1	5|1	6|1	7|1	8|1	9|1
		0|2	1|2	2|2	3|2	4|2	5|2	6|2	7|2
				0|3	1|3	2|3	3|3	4|3	5|3
						0|4	1|4	2|4	3|4
								0|5	1|5
Sum of $T+2L$	3	4	5	6	7	8	9	10	11
Sounds	*sol*$^\#$ *1+*	*la 2+*	*sol 3−*	*re 4−*	*sol*$^\#$ *4−*	*do*$^\#$ *5−*	*fa 5+*	*la 5−*	*do 6*

etc.

The sounds do not correspond to the squares of these sums, as one would presume. Since each interval is greater, we must rather regard them as an enlargement of the natural series of numbers 1, 2, 3, 4, etc. to which they conform when the size is very small.

The figures which give the same sound are ordinarily represented by the distortions which can pass, more pronounced, from one figure to another. Those where several transverse lines are intersected by longitudinal lines appear often in such a way that the ends of the transverse lines are more converging at one side and more diverging at the other. The same attributes of the figures are also noted in the ratios of the axes as 8:3, 11:3, etc.

141. When *the ratios of the axes are as 2 to 1*, the sounds of the same plate will be:

		Number of transversal lines					
		0	1	2	3	4	5
Number of longitudinal lines	0			$re^{\#}\ 1+$	$sol\ 2+$	$fa^{\#}\ 4$	$re^{\#}\ 4$
	1		$si\ 1+$	$do\ 3$	$la\ 3+$	$mi\ 4$	$la\ 4$
	2	$re^{\#}\ 3+$	$do\ 4-$	$fa^{\#}\ 4-$	$si\ 4-$	$mi\ 5-$	$sol^{\#}\ 5$
	3	$sol\ 4+$	$do\ 5+$	$fa\ 5-$	$sol^{\#}\ 5$		
	4	$fa^{\#}\ 5$	$si^{b}\ 5-$	$do\ 6+$			
	5	$re\ 6$					

Here, the series of sounds which conform to the vibrations in which there are only longitudinal lines, 0|2, 0|3, 0|4, etc., is the same as that of the sounds in which there are only transverse lines 2|0, 3|0, 4|0, etc. But the sounds are more acute by two octaves; consequently, they are in this case as the inverse squares of those dimensions to which they relate; in other cases, the ratios are not the same.

142. In the second case, in which the sounds of all the figures (except those where there are no longitudinal lines) contribute to the formation of a single series (Par. 135), and the one where the ratio of one axis to the other is as 8 to 3; this coincidence always takes place a degree later than in the ratio of axes as 5 to 3 (Par. 140). It should be supposed that $n = 3$; each longitudinal line will therefore be equivalent to three transversals, and all the figures in which the sum of $T + 3L$ is the same will give the same sound. Here are the sounds of the same plate:

		Number of transversal lines									
		0	1	2	3	4	5	6	7	8	9
Number of longitudinal lines	0			$re^{\#}\ 1+$	$sol\ 2+$	$fa^{\#}\ 3$	$re\ 4+$	$sol\ 4$			
	1		$re\ 2$	$re\ 3+$	$si\ 3-$	$fa\ 4$	$si^{b}\ 4+$	$re^{\#}\ 5$	$sol\ 5$	$si\ 5-$	$re\ 6$
	2	$si\ 3-$	$fa\ 4$	$si^{b}\ 4+$	$re^{\#}\ 5$	$sol\ 5$	$si\ 5-$	$re\ 6$			
	3	$re^{\#}\ 5$	$sol\ 5$	$si\ 5-$	$re\ 6$						
	4	$re\ 6$									

The figures arranged here vertically give the same sound:

	1|1	2|1	3|1	4|1	5|1	6|1	7|1	8|1	9|1
			0|2	1|2	2|2	3|2	4|2	5|2	6|2
						0|3	1|3	2|3	3|3
									0|4
Sum of $T+3L$:	4	5	6	7	8	9	10	11	12
Sounds:	re 2	re 3+	si 3−	fa 4	si♭ 4+	re# 5	sol 5	si 5−	re 6

etc.

The more the plate is reduced in size, the more these sounds approach those of the natural series of numbers 1, 2, 3, 4, etc.

143. The sounds of the same plate, in *the ratios of one axis to another as 1 is to* $\frac{1}{3}$, were:

		Number of transversal lines					
		0	1	2	3	4	5
Number of longitudinal lines	0			re# 1+	sol 2+	fa# 3	re# 4−
	1		fa# 2	fa# 3	re 4−	sol# 4	do# 5+
	2	mi 4	si♭ 4−	re 5−	fa# 5+	si♭ 5	
	3	sol# 5	si 5−	do# +	mi 6−		
	4	sol 6					

It should be noted here that the sounds of the vibrations in which there are only transverse lines are about three octaves more acute than those in which there are only longitudinal lines.

144. When the ratio of *the major axis to the minor is as 11 to 3*, the sounds of all the modes of vibration, where there are longitudinal lines, form a single series (Par. 135); but the coincidence is carried out at a degree later than in the ratio of 8:3 (Par. 142), and by two degrees later in the ratio 5:3 (Par. 140). It must be supposed here that $n=4$, and the effect of each longitudinal line is the quadruple of a transverse line. All the figures in which the quantity $T+4L$ is the same (Par. 135) give the same sound. The same plate could give the following sounds:

		0	1	2	3	4	5	6	7	8	9	10	11	12
						Number of transversal lines								
Number of longitudinal lines	0			$re^\#$ 1+	sol 2+	sol 3−	$re^\#$ 4	la 4+	$re^\#$ 5	$sol^\#$ 5				
	1		la 2	la 3	mi 4+	si^b 4+	$re^\#$ 5+	$sol^\#$ 5−	do 6	$re^\#$ 6	$fa^\#$ 6	$sol^\#$ 6+	si 6	$do^\#$ 7
	2	si^b 4+	$re^\#$ 5+	$sol^\#$ 5−	do 6	$re^\#$ 6	$fa^\#$ 6	$sol^\#$ 6+	si 6	$do^\#$ 7				
	3	$re^\#$ 6	$fa^\#$ 6	$sol^\#$ 6+	si 6	$do^\#$ 7	$re^\#$ 7							
	4	$do^\#$ 7	$re^\#$ 7											

I will give here the modes of vibration that give the same sound, one under the other:

	1|1	2|1	3|1	4|1	5|1	6|1
				0|2	1|2	2|2
Sum of $T+4L$	5	6	7	8	9	10
Sounds	la 2	la 3	mi 4+	si^b 4+	$re^\#$ 5+	$sol^\#$ 5−

	7|1	8|1	9|1	10|1	11|1	12|1
	3|2	4|2	5|2	6|2	7|2	8|2
			0|3	1|3	2|3	3|3	4|3
						0|4
Sum of $T+4L$	11	12	13	14	15	16
Sounds	do 6	$re^\#$ 6	$fa^\#$ 6	$sol^\#$ 6	si 6	$do^\#$ 7

145. *The ratios of the axes being as 1 to ¼, the sounds of the same plate are the following:*

		0	1	2	3	4	5
				Number of transversal lines			
Number of longitudinal lines	0			$re^\#$ 1+	sol 2+	$fa^\#$ 3	$re^\#$ 4−
	1		si^b...si 2	si^b...si 3	$fa^\#$ 4−	do 5	fa 5−
	2	$do^\#$...re 5	$fa^\#$ 5	si^b 5	$do^\#$ 6+	mi 6+	sol 6
	3	$fa^\#$ 6	la 6−	si 6	$do^\#$ 7−		
	4	fa 7−	$fa^\#$ 7+				

146. When the *size of an elliptical plate is reduced still further*, the sound of the first series, in which there are no transverse lines, will not change very much. The greatest increase to which the first sound will be susceptible does not exceed a semitone, and the sounds of all the other figures in which there are longitudinal lines will become more acute and in the ratios of axes, considered in Par. 135. All these sounds will form a single series: for the ratio of 14:3, $n = 5$; for the one of 17:3, $n = 7$, for the one of 20:3, $n = 6$, and so on. By regarding the quantities that give the same sound we must take successively $T + 5L$, $T + 6L$, etc. I will add the sounds of plates which have similar ratios by reducing all to the same increase.

In the ratio of 14:3:

	1|1	2|1	3|1	4|1	5|1	6|1	7|1	8|1
					0|2	1|2	2|2	3|2
Sum of $T+5L$	6	7	8	9	10	11	12	13
Sounds	$do^{\#}$ 3+	$do^{\#}$ 4+	$sol^{\#}$ 4+	re 5−	$fa^{\#}$ 5	si^{b} 5+	re 6	fa.. $fa^{\#}$ 6

	9|1	10|1	11|1	12|1	13|1	14|1	15|1	
	4|2	5|2	6|2	7|2	8|2	9|2	10|2	
		0|3	1|3	2|3	3|3	4|3	5|3	
							0|4	
Sum of $T+5L$	14	15	16	17	18	19	20	
Sounds	$sol^{\#}$ 6	si^{b}...si 6	$do^{\#}$...re 7	$re^{\#}$...mi 7	$fa^{\#}$ 7−	sol...$sol^{\#}$ 7	la 7	etc.

In the ratio of 17:3:

	1|1	2|1	3|1	4|1	5|1	6|1	7|1
						0|2	1|2
Sum of $T+6L$	7	8	9	10	11	12	12
Sounds	mi 3+	mi 4+	si 4+	mi 5+	$sol^{\#}$...la 5	do...$do^{\#}$ 6	mi 6−

	8|1	9|1	10|1	11|1	12|1	13|1	14|1	
	2|2	3|2	4|2	5|2	6|2	7|2	8|2	
					0|3	1|3	2|3	
Sum of $T+6L$	14	15	16	17	18	19	20	
Sounds	sol 6	si^{b} 6	do...$do^{\#}$ 7	$re^{\#}$ 7	fa 7	sol 7−	$sol^{\#}$...la	etc.

In the ratio of 20:3:

	1|1	2|1	3|1	4|1	5|1	6|1	7|1	
							0|3	
Sum of $T+7L$	8	9	10	11	12	13	14	
Sounds	sol 3	sol 4	re 5	sol 5	si 5	re# 6−	fa# 6	

	8|1	9|1	10|1	11|1	12|1	13|1	14|1	
	1|2	2|2	3|2	4|2	5|2	6|2	7|2	
							0|3	
Sum of $T+7L$	15	16	17	18	19	20	21	
Sounds	la 6	do 7−	re 7+	mi 7+	fa# 7+	sol# 7	la. . .si♭ 7	etc.

I do not want to spend any further time on these experiments (for which the execution and the editing were very difficult) because they are sufficient for judging the passage of more elongated ellipses of a rod or narrow blade. The series 2|0, 3|0, 4|0, etc. represents the transversal vibrations and that of 1|1, 2|1, 3|1, 4|1, etc. the torsional vibrations. I have not added the sound of the only mode of vibration that I have not produced. I do not believe that the differences of the truth, caused perhaps by the small irregularities of the plates, especially when the ellipses are elongated, and by the difficulty of appreciating the very sharp sounds, could equal or exceed a semi-tone.

147. Summary of Research on Elliptical Plates

The following are the results of the research on the sounds of elliptical plates of different ratios of axes:

1. By reducing an axis by a small amount, the first series of vibrations, in which there are no transverse lines, is separated successively to begin a second series of graver sounds, where there are transverse lines intersected by a longitudinal line. These two series are the same in a round plate where the axes are equal and the position of these is of no importance. The intervals of the sounds of the first series, 2|0, 3|0, 4|0, etc., which in a round plate correspond to the squares of 2, 3, 4, etc., are slightly enlarged when the ellipse becomes more elongated, in such a way that they approach more and more the ratios of the squares of 3, 5, 7, 9, etc., which are consistent with the transversal vibrations of a free rod or strip (Par. 71). The absolute pitch of the sounds, which depends on the length, increases by no more than a major third for the first sound when the size of a round plate has undergone a greater diminution.
2. The intervals of sound of the second series, 1|1, 2|1, 3|1, etc., which in the beginning were the same as those of the first series, diminish, little by little, when the size of the ellipse is itself reduced more and more. In this way, these intervals

pass successively to the natural series of numbers 1, 2, 3, 4, etc., which is consistent with that during vibrations of a rod whose motions do not differ essentially. When the ellipse is strongly elongated, for example, in the ratio of axes 17:3 or 20:3, it seems to me that the first sounds of this series approach somewhat more rapidly than the natural series of numbers.

The first sound of this series 1|1 is always approximately in the inverse ratio of the minor axis.

3. The sounds of vibrations where there are only longitudinal lines (counting an elliptical line as two lines, 0|2, 0|3, 0|4, etc.) have between them about the same ratios as those when there are only transverse lines, 2|0, 3|0, 4|0, etc., but they are more acute if the minor axis is more diminished. When the difference of the two axes is slightly greater than 5:4, and slightly less than 4:3, the sounds of the series 0|2, 0|3, 0|4, etc. are more acute by an octave than those of the series 2|0, 3|0, 4|0, etc. When the axes are between them as 1 is to $\frac{1}{2}$, they differ by two octaves; when the difference of the axes is slightly less than 1 to $\frac{1}{3}$, the sounds are more acute by three octaves; when the difference of the axes exceeds slightly the ratio of 1 to $\frac{1}{4}$, the sounds differ by four octaves.

4. In the ratios of the axes 5:3, 8:3, 11:3, 14:3, etc., all the sounds of the vibrations, where there are longitudinal lines, join in order to form a single series. If one counts in the ratio of the axes 5:3 the effect of a longitudinal line as double that of transverse line, in the ratio 8:3 as triple, in the ratio 11:3 as quadruple, etc., all the figures in which the sum is the same give the same sound.

5. If the ratio for the axes is 3:2, the sounds of vibrations in which there is an elliptical line (or two longitudinal lines), 0|2, 1|2, 2|2, 3|2, etc. are the same as those of the vibrations in which there are only transverse lines: 3|0, 4|0, 5|0, etc.

F. Vibrations of Hexagonal Plates

148. They Differ Little from Those of a Round Plate

The figures of the nodal lines in the vibrations of a hexagonal plate resemble those of a round plate by relating to a certain number of diametral and circular lines. But, not knowing how to determine some figures in this way with sufficient accuracy, I prefer to arrange from the lowest to the highest sounds. For greater precision, I will write the figures which can be determined in the same manner for a round plate, separating by a small vertical line the first number, which expresses the diametral lines, from the second, written in Roman numerals, which expresses the circular lines.

149. Figures and Ratios of the Sounds

Of all the figures that can be produced on a hexagonal plate, the one in which two nodal lines intersect, or $2|0$, give the deepest sound. The figures can be shown to be regular as in Fig. 127, but, by small changes in the points of touching and rubbing, the position of the lines can be changed without alteration of the sound, in such a way that their direction no longer has a certain ratio determined by the shape of the plate. I will regard this sound to be the deepest as *do 2*, in order to compare it with other sounds.

The sound of $0|1$ (Fig. 128) is almost a minor seventh more acute than the preceding sound; it will therefore be si^b *2*.

In $3|0$, the nodal lines can terminate in the middle of the sides (Fig. 129) or at the corners (Fig. 130); in the first case the sound will be *re 3*, in the other *fa 3*.

In $1|I$, a diametral line that cuts the circular line can pass through the middle from one side to the other (Fig. 131), or at an angle to the other (Fig. 132), or in every other direction, without changing its sound, which will be more than two octaves higher than the first (*do 4*).

$4|0$, which gives $do^\#$ *4*, ordinarily shows itself as Fig. 133, but the direction of the lines is arbitrary.

Figure 134 appears to be a distortion of $5|0$, the sound is between $sol^\#$ *4* and *la 4*.

Figure 135 represents $2|I$ and Fig. 136 represents $0|II$, in which the inflexions of the other circle are found at the corners to give the same sound *si 4*. These two figures can pass from one into the other through intermediate distortions.

Figure 137 is equal to $6|0$, and Fig. 138 gives the same sound *re 5*. I will not decide if Fig. 138 is a distortion of Fig. 137, or $3|I$, with the inflexions in the middle of the sides.

It appears that Figs. 139 and 140, whose sound is $re^\#$, represent $0|II$ in another way than Fig. 136, the inflexions of the outer circle being in the middle of the sides.

$1|II$ is shown in two different ways: the diametral line can terminate at the corners (Fig. 141), or in the middle of the two sides (Fig. 143). In the first case, the sound will be *fa 5*, in the other, *la 5*. Figure 143 is often distorted into Fig. 144.

$3|I$ with the lines that terminate at the corners (Fig. 142) gives $fa^\#$ *5*.

I will not decide if Figs. 145 and 146, which do not differ essentially, represent $8|0$ or $4|I$; the sound is slightly more acute than *si 3*.

$2|II$ (Fig. 147), of which Fig. 148 is a distortion, gives *re 6*.

Figure 149, which gives *mi 6*, appears to represent $9|0$.

Figure 150 is perhaps $3|II$, with the diametral lines that terminate in the middle of each side; the sound is *fa 6*.

Figure 151, which shows itself sometimes as Figs. 152 and 153, represents $3|II$, in the manner that the diametral lines are terminated at the corners; the sound is *sol 6*.

The same sound is produced by $6|I$ (Fig. 154), which is often transformed to Fig. 155, and through Fig. 156, which I do not rank with *certitude*.

I will not press the production of figures and sounds of hexagonal planes much further. The sounds of the modes of vibration $2|0$, $3|0$ (when the lines terminate in the middle of the sides (Fig. 129)), $4|0$, $5|0$, etc. seem to be in the ratios of the squares of the numbers 2, 3, 4, 5, etc., the same as on a round plate. The sounds of the figures in which there is a circular line, $0|I$, $1|I$, $2|I$, $3|I$, etc., when the lines are terminated at the corners, are also approximately in the ratios of the squares of 2, 3, 4, 5, etc. In the figures where there are only one or two circular lines, etc., the sounds seemed to be as the squares of 1, 2, 3, etc. All this is almost as in the round plates, but if the nodal lines do not terminate at the corners, the sound is ordinarily deeper than in the opposite case, because, in the first case, the vibrations are slowed down by the peaks of the angles.

Very few of the figures of a hexagonal plate have the attributes that are necessary for formation of regular patterns, when one composes several plates on which is found the same figure, as I have shown in the square plate (Par. 110).

To avoid verbosity, I have not commented on the manner of producing each figure which I have set forth in general in Par. 92, and in the remarks on the modes of producing various sounds of other plates. Here, I have only described the most suitable points to clamp the plate and apply the bow.

G. Vibrations of Semicircular Plates

150. The figures Are Half Those of a Round Plate

In all the modes of vibration of a semicircular plate, the figures of the nodal lines are related to a certain number of semi-diametral and semicircular lines. Most of the figures, especially those in which there are semicircular lines, appear in such a way that in composing two similar figures on plates of equal size, approximately the same figures are found as can be produced on a round plate.

151. Ratios of the Sounds

The ratios of the sounds of a semicircular plate will be the following if I regard the gravest sound in Fig. 209 as *do 2*:

		Number of semi-diametral lines							
		0	1	2	3	4	5	6	7
Number of semi-circular lines	0				Fig. 204 *fa 2*	Fig. 205 *re# 3+*	Fig. 206 *do 4−*	Fig. 207 *fa# 4*	Fig. 208 *si 4*
	1	Fig. 209 *do 2*	Fig. 210 *re# 3+*	Fig. 211 *do 4+*	Fig. 212 *sol# 4*	Fig. 213 *re 4−*	*sol 5*	*si 5*	
	2	Fig. 214 *re 4+*	Fig. 215 *si♭ 4*	Fig. 216 *mi 5*	Fig. 217 *la 5*	Fig. 218 *do# 6+*	*fa 6*		
	3	*fa 5*	*si♭ 5*	*re 6+*	*fa# 6*				
	4	*re# 6*	*sol 6*						

I would have presumed that one could also produce a mode of vibration in which there were two semi-diametral lines, almost as Figs. 211 or 216, if there were semicircular lines; but this has not been successful.

The sounds of the modes of vibration in which there are only semi-diametral lines (Figs. 204–208) do not differ much from the squares of the numbers 3, 4, 5, 6, etc. In regarding this series as isolated from the others, all the sounds of the modes of vibration in which there are semicircular lines approach the ratios of the squares of $D + 2C$, where D is the number of semi-diametral lines and C the number of semicircular ones, neglecting the enlargement of the intervals due to the preponderance of semicircular lines and their diminution due to the preponderance of semi-diametral lines. Everything that takes place here is approximately the same as on a round plate.

152. Vibrations of Plates That Are a Small Part of a Round Plate

If the shape of the plate is a quarter or a sixth or in general a part of a round plate, many of the figures appear in such a way that they are part of those that can be produced on a round plate, and are related to the lines in the diametral or circular directions.

H. Vibrations of Triangular and Other Plates

153. Vibrations and Sounds of an Equilateral Triangular Plate

Some figures of nodal lines of an equilateral triangular plate could be arranged according to the number of lines or parallels or normals to the base; but, because several figures do not want to be accommodated to this manner of viewing them, I range here the figures that I have observed according to the lowness or highness of pitch of their sounds.

The figure that gives the deepest sound is Fig. 219 that can also appear as Fig. 220. I will attribute to this figure the sound *do 2* to compare it with the others. Figure 223, which can transform itself easily from Fig. 222, gives a slightly more acute sound, $do^{\#}$ 2. But Figs. 219 and 223 are surpassed by Figs. 220–222. The sound becomes more acute when the figure approaches Fig. 223, and graver when the figure approaches Fig. 219. The biggest difference surpasses slightly more than a semi-tone.

The sound of Figs. 224 and 225, which are essentially no different, will be slightly more acute than *re*# *4*;

That of Fig. 226, is close to *fa 4*;

That of Fig. 227, *la 4*;

.... Fig. 228, which can also show itself as Fig. 229 or 230, *re 5;*

.... Fig. 231 and the distortions, Figs. 252 and 253, *re*# *5;*

.... Figs. 234 and 255, *sol 5;*

.... Figs. 236–238, which are all variations of each one another, and of Fig. 259, *si*♭ *5;*

.... Figs. 240–242, *re*# *6;*

.... Fig. 243, *sol 6.*

154. Compositions and Portions of Figures That Can be Produced on Several Other Plates

All the figures of equilateral triangular plates, when several plates with the same figures are combined, form more or less complex regular designs.

Several complex figures are also formed by composing four plates in such a way that they form a larger triangle, as in Fig. 244; for example, Fig. 229, which is formed by combining Fig. 219 four times, and Fig. 243, which contains Fig. 226 four times. One would have the same Fig. 237 if one surrounded a plate on which is found Fig. 226, with three others on which one has produced Fig. 223. Some parts of an equilateral triangle give several figures that can be regarded as parts of those that form an equal triangle but the ratios of these sounds are different. On a plate whose shape is a trapezoid, which one has produced by cutting the fourth part of an equilateral triangle (Fig. 245), the figures are almost as those on triangular plates, except the part of the figure which should be found on the intersected part. Some plates whose shape is a right triangle produced by cutting vertically an equilateral triangle (Fig. 246) also give several figures which are half of those of an equilateral triangle. Several figures that can be produced by composing two triangular plates can also be produced on a rhomboidal plate of the same shape (Fig. 247). Very few of the figures on a hexagonal plate result from the composition of triangular plates; the only ones I know are those of:

Figure 136, which is formed by taking Fig. 219 six times,

Figure 139 222,

Figure 150. . . . . 237.

I. Remarks on Some Practical Usage of Plates

155. On Two Chinese Instruments

With us, as far as I know, plates are not used for music, except for carillons, which consist of rectangular blades of glass, steel, or other sonorous material, beaten with sticks, or by small hammers put in motion by a keyboard. The mode of vibration is represented in Figs. 47 and 24. In China, a musical instrument is used, called a *king*, which consists of metal or slate plates, and has almost the shape of a bracket, as in Fig. 248. The ratios of the dimensions are: $cd = 2$, $be = 3$, $ab = 6$, $ac = 9$. A line is drawn parallel to *ca*, at a distance of half of *cd,* and another parallel to *ab* at a distance of half of *eb*; in the place *n,* where these two lines intersect, a hole is drilled, to which one plate is suspended. It is hit with sticks at the place marked by *g.* More information is found in the *Mémoires concernant les Chinois,* vol. VI, written by Amiot,[4] p. 2, p. 255, and in an added dissertation: *Essai sur les pierres sonores,* also in Vol. XIII of the *Histoire générale de la Chine,* translated by Grosier, pp. 300 and 772. Those experiments which I made on glass plates of the same shape showed that the nodal lines were as in Fig. 249; consequently, the points where the plate is suspended and beaten are the most suitable.

Another Chinese instrument, which finds its place here, is called *gongong* or *tamtam.* Those that I saw, especially in Copenhagen, were of copper cast in a single piece, in the shape of a tambourine, about six decimeters in diameter. The edge was a little more than 6 cm high, and it was the thickness of a finger. On the sounding plate in the middle, of less thickness, impressions of the blows of a very strong hammer can be seen, which have served to increase considerably the elasticity of the plate by the resistance of the edge against the tendency of the plate to expand.[5] Consequently, the manner of elasticity of this plate is completely the opposite of that of a kettledrum, where it is made by tension. One hits the area with a stick whose end is enveloped in a cloth or something else soft; the sound is extremely strong and resonant, and is accompanied by a slight upturn that produces an alarming effect. This instrument was employed in Copenhagen, with success, to express in an Oratorio the trembling of the earth at the death of Jesus Christ; in China it is used to give signals.

---

[4] Jean-Joseph-Marie Amiot (1718–1793). French Jesuit missionary sent to China.—*MAB*

[5] I have been told that the inequalities on the outside of the *gongong* or *tamtam* appear to be produced by impressions of the pumice that was used in the mold. Some experiments have shown that the pieces of a similar instrument are not malleable.—*EFF Chladni*

Section 8: Vibrations of Bells and Vessels

156. General Remarks

The vibrations of a bell or of a round vessel correspond to those of a round plate in which there are only diametric nodal lines (Par. 126). These sounding bodies can be divided into four, six, eight, or, in general, into an even number of vibrating parts, separated by nodal lines that intersect in the middle, where the neck of the bell is. The principal difference from a round plate is the fact that the curves caused by the vibrations do not apply to the straight directions, but to the curves that already exist in the shape of the sounding body.

157. Manner of Producing Vibrations and Making Them Visible

When a bell is struck, one hears the sound most strongly. But, by listening with attention, one will often find it accompanied by a confused mixture of more acute sounds that are not very harmonic. However, one can separately produce each sound to which the bell is susceptible, by touching with the fingers, or in some other way, one or more of the nodal lines of the mode of vibration that one wishes to produce and applying the bow to the middle of one vibrating part. One cannot make use of sand to render visible the nature of the vibrations, since there is not a flat surface. It is therefore necessary to put some water in the bell or in the vessel, which, according to whether one produces the first or the second or another mode of vibration, is driven back by four, six, or more vibrating parts of such sort that the excitations of the surface are visible, as is shown in Fig. 252 or 257. The same excitations are seen outside if the bell is surrounded by water. When one spreads a little very dry lycopodium powder on the surface of the water, the division into four, six, or a larger number of parts is made visible by the more enduring figures.

© Springer International Publishing Switzerland 2015
E.F.F. Chladni, R.T. Beyer, *Treatise on Acoustics*,
DOI 10.1007/978-3-319-20361-4_11

158. The Fundamental Sound of a Bell

The simplest mode of vibration, which yields the deepest sound, can be produced, without mixing other sounds, if one touches the bell or vessel with the tips of one's fingers, in two points either opposite or far removed from one another by a quarter of a circle, applying the bow at a distance of 45° from a nodal line whose position is determined by the contact. For example, if the bell (Fig. 250) is touched at *m* or at *n* or at the point of the line *pq*, one must apply the bow in the direction *cf* or *hg*. The four parts *qfn, ngp, pcm, mhq* make their motions, as we have shown on a round plate, in such a way that the two opposing sides are bent inwards, while the other two are bent outwards; the lines *mn* and *pq* remain immobile. The bell takes alternately the curves represented in Figs. 251a, b. If one puts some water in the bell, the excitations are seen on the surface as in Fig. 252.

159. Application to a Harmonica Bell

A harmonica bell which turns around the axis and whose vibrations are produced by rubbing with a wet finger or with other suitable material, and a round glass vessel rubbed in the same way not far from the edge in the direction of the periphery, also divide into four vibrating parts. But the position of these parts changes at every instant. The mode of vibration and the sound are the same as if one struck the bell or if one applied the bow of a violin to it, but the location where the motion is produced has another relation to the position of the nodal lines and the vibrating parts. When the motion is produced by striking or by applying the bow in a diametrical direction, the point is approximately in the middle of a vibrating part and the nodal lines are found at a distance of 45°. But when motion is produced by striking in the direction of the periphery, a nodal line passes through the point of striking and the part of the bell (Fig. 253) where the striking is done in the direction *mn* takes the positions *fg* and *pq* alternately.

We cannot touch the harmonica bell in more than one point at the same time without avoiding the vibrations except in points that are opposite or distant from each other by a quarter of the periphery.

160. In Irregular Bells the Sound Is Not the Same Everywhere

The construction of a harmonica is often rendered laborious by the inequalities of the sound of the same bell when it is struck in different points. The inequalities of the sound can be caused by irregularities in the thickness or by some eccentricities; the sound is slightly different if a nodal line passes through the abnormal point, or if this point is found in a vibrating part. An experiment on a porcelain cup with a handle can show this. If one applies a violin bow to the point where the handle is found, or opposite it, or at a distance of 90°, the position of the nodal lines will be as in Fig. 254. The sound will be graver than if the motion were produced at a distance of 45 or 135° because in this case the handle does not undergo vibrations, the position of the nodal line being as in Fig. 255.

One therefore understands that by applying the bow successively to all the places on the periphery, it produces eight alternate variations of a deeper sound and a higher one, as in each rotation of an abnormal bell of a harmonica. A bell for sounding which has the same defect would, however, emit a pure sound when struck at a point where one of these two sounds is predominant, and when the position of the nodal lines is fixed by a type of damping, applied at a distance of 45 or 135° from the point struck.

161. Other Types of Vibrations

In the second type of vibration, a bell or a round vessel divides into six vibrating parts, like the round plate in Fig. 100. To produce this sound, we apply the bow at a distance of 90° from the point where a nodal line has been fixed by touching. We can also touch (at the same time) two points distant from one another by 60°. The bell bends alternately to the curves shown in Figs. 256a, b. When water is put in the bell, the excitations of the surface appear as in Fig. 257.

The same type of vibrations are produced by touching together two points separated from one another by the eighth part of the periphery and applying the bow to the middle between these two touched points, or in another suitable place. In the other modes of vibration, the bell or the vessel can be divided into 10, 12, or more vibrating parts, as many as its size and thickness permit.

If the shape of a bell or vessel is even enough, and the thickness is the same all the way through, the series of possible sounds is like the squares of 2, 3, 4, etc. When the gravest sound is *do 2*, the series of possible sounds will be:

Number of vibrating parts:	4	6	8	10	12
Sounds:	*do 2*	*re 3*	*do 4*	*sol$^{\#}$ 4-*	*re 5-*
Numbers whose squares correspond to these sounds:	2	3	4	5	6
	etc.				

This series will be that of a hemispherical harmonica bell, or of another similar vessel. If the shape is different, or if the thickness is not the same towards the edge and towards the middle, all the intervals could become lesser or greater, in the way that the distance between the first and second sounds could be less than an octave, or greater than a twelfth, and in the same way the other distances become greater or lesser. However, it is necessary to look at the cited series as the average term for the distances between one sound and another that are the same as on a round plate divided in the same way.

L. Euler (*de sono campanarum in Nov. Comment. Acad. Petrop.* vol. X) claims that the series of possible sounds of a bell is like 1, $\sqrt{6}$, $\sqrt{20}$, $\sqrt{50}$, $\sqrt{105}$, $\sqrt{196}$, etc. Golovin, having applied Euler's research on the vibrations of the rings of harmonica bells, found that if the fundamental sound of a bell, divided into four vibrating parts, is equal to 1, the other sounds should be like the squares of 2, 3, 4, 5, etc. But these results are not observed in the experiment, and the assumptions on which this research was founded do not conform to nature.

One must not aim to explain a bell's vibrations by the vibrations of its rings; the motion of the vibrating parts of a ring and their link to the sounds are quite different from those of a bell, and from those that result from Euler's and Golovin's research. It seems to me that there will be no other way of determining, theoretically, the vibrations of a bell, assuming that the true expressions for the vibrations of a round plate in Figs. 99, 100, 101a, and 102a have been found, than to apply the curvature of a straight edge to the curvature that already exists from the shape of the sounding body.

162. Laws for the Absolute Frequency of Vibrations

If n represents the number whose square corresponds to each mode of vibration, D the thickness, L the diameter, R the stiffness, and G the specific gravity, the sounds of vessels or bells, whose shape is the same, will be $\frac{n^2 D}{L^2} \sqrt{\frac{R}{G}}$, as in other rigid bodies (Par. 75).

163. The Vibrations of Sounding Bodies of Other Shapes Are Still Not Known

The vibrations of other rigid membranous bodies, for example, of a spherical shape, cylindrical shape, etc., are quite unknown, and it will be very difficult to determine them by experiments, and even more difficult to do so theoretically.

Section 9: On the Coexistence of Several Modes of Vibration in the Same Sounding Body

164. Several Vibrations Can Coexist

Many or all of the modes of vibration that can be produced separately can coexist in the same sounding body; therefore, by listening with enough attention, one hears the sounds that correspond to all of these types of vibration. To get an idea of the changes in the shape of the elastic body, one must not apply the curvature that corresponds to a mode of vibration to the original shape of the body, but to the curvatures that already exist in each moment due to the other modes of vibration. Such a coexistence of many types of vibrations and many sounds is not necessary, as some have claimed, because in every mode of vibration where there are nodes, one can, by touching them or by applying a mute, exclude all types of vibrations in which these points should be moving, and produce the desired motion, and the sound that corresponds to it, without mixing in others.

165–170. Coexistence of Several Vibrations in a Single String

165. In the simplest transverse motion of a string, this coexistence of many sounds is fairly well-known. While the entire string vibrates, each half, each third, and in general each aliquot part can vibrate as well; therefore, one hears, in addition to the fundamental sound equal to the unit, the sounds that correspond to the numbers 2, 3, 4, 5, etc. The curvature that corresponds to a mode of vibration must therefore be applied to the curvatures in which the string bends in each instant due to the other modes of vibration. To explain these combinations of many curvatures, I will use some examples borrowed from the work of Count Giordano Riccati, *Delle corde ovvero fibre elastiche* (*On strings of elastic fibers*).

© Springer International Publishing Switzerland 2015
E.F.F. Chladni, R.T. Beyer, *Treatise on Acoustics*,
DOI 10.1007/978-3-319-20361-4_12

166. To imagine the combination of the two curves (Fig. 5, *BDF2DA* and *BGC2GA*), one of which pertains to the simplest vibrations of the entire string (Fig. 1), and the other to those of the same string divided into two halves, one must, for a point *H*, extend the ordinate *HD* towards *E*; make *DE* = *DH*; and make the curve *BEF2EA* pass by every point that can be determined in the same manner; the resulting curvature will be that of the string in its first state of rest.

After the fourth part of a vibration of the entire string, each half makes a half-vibration, and in that moment the shape of the string is like *BDF2DA*. After a half-vibration of the entire string, and a vibration of each half, the curvature is similar to *BGC2GA* in an inverted position. When the entire string made $\frac{3}{4}$ of a vibration and each half made $1\frac{1}{2}$ vibrations, the shape is like *Bdf2dA*. Finally, after the entire string made a vibration, and each half made two, the string arrived at the second state of rest, and the shape is *Bef2eA*; which is produced when one makes *de* = *HG*, or when the curve *BE2EFA* is found alternatively here and there on the axis.

These alternative positions of the curvatures in the resting states, or *BEF2EA* = *A2efeB*, take place in all the combinations of vibrations of the entire string with vibrations of the string divided into an even number of parts. In these cases, all the points of the string never pass at the same time by the axis *BCA*, because, when the point F arrives at *C*, the point *E* is advanced beyond point *H* by *HG*, and the point *2E* has not yet arrived at *2H*, having remained behind *2H2G*.

167. In the combinations (Figs. 6 and 7) of the first mode of vibration of a string *BDF2DA* with the third, where the string is divided into three sections *BGS2G2S3GA*, its curvature is susceptible to two different positions, either like Fig. 6, or like Fig. 7. Taking at each point *H* the ordinate *HD* extended as much as necessary; and making *DE* = *HG*; then the point *E*, and all the other points determined in the same manner, will form the curve *BEN2E2N3EA*, which agrees with the string in its first resting state. After a half-vibration of three sections, the shape of the string will be like *BDF2DA*. When the entire string has made a half-vibration, and the three sections have made $1\frac{1}{2}$ vibrations, the string will be in the straight line *BCA*, and all of its points will pass by the axis at the same time, as is also the case in all other combinations of vibrations of the entire string with those of the string divided into an odd number of sections. After the three halves have made $\frac{5}{2}$ vibrations, the string will once again take the form *Bdf2dA*, and finally, after a vibration of the entire string and three vibrations of the string divided into three sections, the string will be in the second resting state, and its shape will be *Ben2e2n3eA*, similar to *BEN2E2N3EA*, as in all the other combinations of the first mode of vibration with those where there is an odd number of sections.

168. Just as by the combination of two curves (Figs. 1 and 2), curves *BEF2EA* and *Bef2eA* (Fig. 5) are formed, which produce a mix of the fundamental sound and its octave; thus, by the combination of these curves with that of the third mode of vibration (Fig. 3), even more new curves are formed, which simultaneously produce the first, second, and third sound. One can apply the one that agrees with the fourth mode of vibration to these new curves, to find the curves that give a mix of sounds corresponding to the numbers 1, 2, 3, and 4. By continuing in the same mode, one can move on to even more complex curves, in which the number of sounds corresponding to the natural progression of the numbers, mixed with the fundamental sound, becomes greater and greater.

If, according to Taylor, Daniel Bernoulli, and Giordano Riccati, for the first type of vibration of a string (Par. 39) $y = A = \sin \frac{\pi x}{L}$, for the second, $B = \sin \frac{2\pi x}{L}$, for the third, $C = \sin \frac{3\pi x}{L}$, the general expression for all of these combinations of curvatures will be $y = A \sin \frac{\pi x}{L} + B \sin \frac{2\pi x}{L} + C \sin \frac{3\pi x}{L}$, etc.; and when the initial curve is expressed using this equation, at the moment when a vibration of the entire string is achieved, y will be equal to $-A \sin \frac{\pi x}{L} + B \sin \frac{2\pi x}{L} - C \sin \frac{3\pi x}{L} + D \sin \frac{4\pi x}{L}$, etc. This curve is the same as the original, in an inverted position: x represents here any abscissa with y as its ordinate, L the length of the string, π the half-circumference of the circle whose radius is equal to 1. The coefficients A, B, C, D, etc. that can be taken as positives or negatives express the greatest oscillations in mid-cycle for each mode of vibration. If, according to Euler and others, a vibrating string is susceptible to even more shapes, which are not included in this equation, that does not prevent the combinations of many types of vibrations.

169. Until now, the only question was of combinations of the fundamental sound and those where the string is divided into aliquot parts; but it is also necessary to mention the case where two types of vibrations of the aliquot parts happen at the same time.

The combination of curves $BDC2DA$ and $BGS2G2S3GH$ (Fig. 8.1), which pertains to the divisions of the string into two or three sections, will form, by making $DE = HG$, the curve $BEN2E2N3EA$, which pertains to the mix of these two sounds. After a vibration of the two halves of the string, each third part makes $1\frac{1}{2}$ vibrations; the shape of the string will thus be $BDC2DA$, taken at the other side of the axis. This is not a resting state, because the string, divided into three sections, continues in its motion. A resting state will only occur after two vibrations of the string divided into two sections, and three vibrations of the string divided into three parts; therefore it will have the curve $Ben2e2n3eA$ (Fig. 8.2). Figure 8.1 is separate from Fig. 8.2 in order to better distinguish the curvatures.

For the string to get from one resting state to the other, it is always necessary to achieve two vibrations of the halves and three vibrations of the third parts. In the time necessary for this effect, the entire string will have made one vibration. In this combination, and in all other combinations of vibrations of aliquot parts, one will therefore always hear the fundamental sound, which pertains to the unit, when the numbers of the aliquot parts are expressed in the lesser terms.

170. There is no way to prevent the coexistence of acute sounds with the fundamental sound; but each sound of the string, divided into aliquot parts, can be produced, without any mix of other sounds, by touching the nodes to exclude all types of vibrations in which these points should be moving. It seems that the reason the harmonic sounds of a cello or a violin are softer than the same sounds produced in the regular manner is mainly due to these sounds not mixing with others.

171. Coexistence of Several Vibrations in an Organ Pipe

An organ pipe, or another wind instrument, can also produce two sounds at the same time, when the manner of breathing is in between those that serve to produce one sound or the other.

Similarly, one type of longitudinal vibration of the air is not prevented by another. The same thing happens in the longitudinal vibrations of strings and rods.

172. Coexistence of Several Vibrations in a Rod

A rod or band that makes transverse vibrations will never simultaneously produce the sounds contained in the natural series of numbers, but rather sounds that are not very harmonious, which can only be expressed by the squares of certain numbers (Pars. 69–74). When, for example, one of the ends is fixed and the other is free (Par. 69), the sounds that can coexist correspond to the numbers 36, 225, 625, 1225, 2025, etc.; or if one looks at the fundamental sound as a unit: $1, 6\frac{1}{4}, 17\frac{13}{36}, 34\frac{1}{36}, 56\frac{1}{4}$, etc. The vibrations will therefore never coincide at the same time.

173. Coexistence of Several Vibrations in a Fork or a Ring

In the fundamental sound of a fork (Par. 88), one cannot prevent the coexistence of other sounds, because one cannot touch the middle, which is at rest in all the modes of vibration. The sounds that one will be able to hear at the same time, by looking at the fundamental sound as a unit, are $1, 6\frac{1}{4}, 11\frac{1}{9}, 17\frac{13}{36}, 25, 34\frac{1}{36}$, etc.; or in whole numbers, 36, 225, 400, 625, 900, 1225, etc.; consequently the vibrations will not come back together until after the 36th vibration of the two branches of the fork. Nevertheless, the sound of a fork (for example a tuning fork) would be agreeable, because the coexistence of other sounds is almost imperceptible due to their great distance from the fundamental sound. Each mode of vibration, in which other modes can be excluded, can be produced, without mixing in the others, by touching the nodes. A ring-shaped object will give several sounds at the same time. For example, when it is suspended by a wire and struck, the sounds will conform to the squares of 3, 5, 7, 9, etc. (Par. 89). By touching the nodes, the mixing of sounds can be prevented.

174. Coexistence of Several Vibrations in a Plate

Plates of any shape are susceptible to many motions at the same time, so one hears all the sounds that correspond to these motions. By touching and hitting a plate at different points, one often hears more than one sound at the same time. The same thing also happens sometimes when one uses a violin bow, and in that case one cannot produce a distinct shape, because the shape that corresponds with one type of motion is destroyed by the other. It is thus necessary to touch another or even several points at the same time, where there is a node for one of these modes of vibration and not the other. The easiest way to discern such a mix of sounds is by holding a round plate in the middle and hitting it, or applying the bow, without fixing the position of the nodal lines in any way. One will hear many sounds, having prevented the other motions, and there will never be a distinct shape.

175. Coexistence of Several Vibrations in a Bell

On a bell, the shock not only produces the simplest motion, where it is divided into four vibrating parts. At the same time it can also vibrate, divided into six, eight, or more parts, and one hears, in addition to the fundamental sound, a weak coexistence of sounds which, the fundamental sound being equal to the square of 2, just corresponds to the squares of 3, 4, 5, etc. But this coexistence can be prevented by touching the bell with mutes applied to the nodal lines.

176. Authors Consulted

(*Note*: *The Coexistence of Sounds in the Same Sounding Body Is Not the Basis of Harmony*). The best research on the coexistence of several types of vibrations in the same sounding body can be found in some dissertations of Daniel Bernoulli (*Mé m. de l'Acad. de Berlin*, 1753 and 1765, and *Nov. Comment Acad. Petrop.* vols. XV and XIX); in the research on sound by Lagrange (*Miscel. Taurin.* vols. 1 and 2); in the work of Giordano Riccati, *delle corde ovvero fibre elastiche* (Append. in Schediasm IV); in Matthew Young's *Enquiry into the principal phenomena of sounds and musical strings*, p. II. Mersenne already knew of the coexistence of acute sounds with the fundamental sound of a string, but he did not explain it well; Descartes (in Epist Part II, ep. 75 and 106) explained it better, but he attributed this quality exclusively to irregular strings.

Several authors have looked at the coexistence of sounds included in the natural series of numbers (which, according to the true principles, is just an idiosyncratic phenomenon) as an essential quality of each sound, and as the essential difference between a distinct sound and a noise. They have taken this quality to be the basis of all harmony, by believing that an interval is consonant *because* the acute sound can be heard with the fundamental sound. They did not know that if one hears more than one sound at the same time, it is merely a result of the coexistence of several types of vibrations; that in many sounding bodies, the series of possible sounds is very different from the natural series of numbers; and that one can produce each mode of vibration where there are nodes by touching the points or nodal lines that should be moving in other modes of vibration. According to their principles, the perfect minor chord, if one does not want to use sophisms, would not be consonant, and on a harmonica bell, the ninth (4:9) would be the first consonance, because it is the first sound that can mix with the fundamental sound, etc. Daniel Bernoulli and Lagrange (in the cited papers) refuted these false principles sufficiently. It will always be more natural to look at the greatest or least simplicity of numerical relationships to vibrations as the basis of harmony. However, all efforts to develop laws of harmonics, assuming false physical principles, have not been too detrimental to the theory and practice of harmonics. Despite the diversity of principles, all agree that intervals that can be expressed by the numbers 1–6, and by their multiplications by 2, are consonant and all others are dissonant.

177. Coexistence of a Grave Sound When Two or More Acute Sounds Are Produced

The coexistence of a grave sound, when two or more acute sounds are produced, is a general quality which applies to all sounds. The ear perceives not only the effect of the ratios themselves, but also the effect of the concurrences of the vibrations at the same moment; it hears the concurrences, in which the ear is hit by two blows, like a grave sound whose vibrations take place in the same space of time. The grave sound, caused by the concurrences, will thus always be equal to the unit, if one expresses the sounds that are really produced by the smallest whole numbers. I will express here the time intervals in which the vibrations take place by dots. If one produces two sounds that make a fifth, for example, *do 2* and *sol 2*, the following concurrences will happen, which will yield a resonance of sound equal to the unit, *do 1*:

If the major third *do 3*: *mi 3* is produced, there will be the same resonance of the sound *do 1*, equal to the unit:

The minor third *mi 3* and *sol 3* will yield the same result, as well as the major sixth *sol 2* and *mi 3*, etc.

To make this grave sound more perceptible to the ear, it is necessary that the sounds, truly produced, be extended enough and of roughly the same strength; it is also necessary that the ratios of the sounds be very exact or differ little from exactitude.

178. Beats in Poorly Tuned Instruments Are the Same as Coexistence

If the vibrations of two sounds rarely come back together, these concurrences are perceived as beats, very disagreeable to the ear in a badly tuned instrument. The closer the interval gets to exactitude, while tuning the instrument, the more these beats become undetectable, until finally they are lost in the sensation of the weak resonance of a grave sound. An instrument is not well-tuned if any interval produces beats.

Sauveur proposed that we use these beats to find the absolute values of vibrations, by comparing the interval of the sounds of two organ pipes with the interval of the time that passes between two beats. Sarti did similar experiments in the presence of the Imperial Academy in Petersburg, 1796.

179. Authors Consulted

To the best of my knowledge, the first person to mention the coexistence of a grave sound equal to the unit was G. A. Sorge, who, in his *Instruction pour accorder les orgues et les clavecins (Anweisung zur Stimmung der Orgelwerke und des Claviers,* Hamburg, 1744), says roughly (pp. 40 and 41):

> *Where does it come from that if one tunes a fifth 2:3, the third sound is heard in a weak resonance, and always the octave of the sound deeper than the fifth? Nature shows that for 2:3 the unity is still missing, and it wants to have it as well so that the order of 1, 2, 3 is perfected. This is why playing a fifth of three feet on an organ makes the sound more perfect by producing a third sound of almost the same strength as a weak tenor bell; and not only fifths, but also thirds do the same thing, etc.*

Romieu observed this phenomenon and reported it at the Montpelier Academy, 1753. Tartini, to whom this discovery is often attributed, mentioned it in his *Trattato di Musica*, Padova, 1754.

The best remarks on this third sound are found in *Recherches sur le son*, by Lagrange, *(Miscell. Taurin.* vol. 1) Sect. 64, and in Matthew Young's *Enquiry into the principal phenomena of sounds and musical strings*, p. 2, sect. vi.

Tartini claimed that this third sound was more acute than an octave, which it is not really. He regarded this phenomenon, combined with the purported coexistence of the series of sounds 1, 2, 3, 4, 5, etc. in each fundamental sound, as the basis of harmonics. Mercadier de Belesta expertly refuted some of Tartini's false assertions in his *Système de Musique*, Paris, 1776.

Abbé Vogler uses this third sound to substitute for a large organ pipe that produces the same sound in the ordinary way: two small pipes that produce it as the unity for the numbers of their vibrations.

Section 10: On the Coexistence of Vibrations with Other Sorts of Motions

180. These Motions Can Coexist

Vibratory motions can coexist with all other types of motion (Par. 1) in an infinite number of different modes, which was demonstrated by Daniel Bernoulli and L. Euler in vols. XV and XIX of *Nov. Comment. Acad. Petrop.* and established through experimentation. These coexistences of different motions take place in all sounding bodies, without exception. It is possible, for example, to produce the sound of a string stretched over a piece of wood, or that of a blade, a tuning fork, a bell, etc., and while the vibrations continue, to imprint upon this sounding body a motion of rotation around its axis, and at the same time a progressive motion by throwing it. All these movements could happen at the same time, without one being prevented by the other; but the absolute motion of each point would be very complex.

181. On a Very Common Coexistence of a Circular Motion with Vibrations

Movements composed of vibrations and circular motions are often noticeable in strings of a sufficient length. The space that the vibrating string passes through in its oscillations seems half-transparent to us, and its limits especially distinguish it, because the string stays there longer than it stays in the middle of this space. This space seems sometimes to shrink and enlarge alternatively towards one side or the other; sometimes in one half of the string this image is on this side of the axis, while in the other half it is on that side, and sometimes two images seem to get closer to, and then farther away from, each other. The nature of these complex

© Springer International Publishing Switzerland 2015
E.F.F. Chladni, R.T. Beyer, *Treatise on Acoustics*,
DOI 10.1007/978-3-319-20361-4_13

movements is seen most distinctly in a rather elastic rod, for example, an iron wire held in a vise, by making it bulge out enough that the slowness of the motions allows the eyes to follow the progress of the wire. When, after having pulled this wire out of the ordinary position, one lets it go at an oblique angle, the circular motions mix with the vibrations (Fig. 20), because the wire is more set in one direction than in the other, and because it leans at an oblique angle against the jaws of the vise. For the same reason, these composed movements will take place in a string, if the direction in which the motion is produced makes an oblique angle with that of the bridge. The progression of the rod, which does not seem to differ from that of a string in its composed motions, is represented in Fig. 258, by taking all the vibrations of equal size. When the iron wire, whose ordinary position is at the middle of the figure, is pulled towards C and released in the direction CD, the progression of wire will be:

CDxCuDtCsDrCqDpCoDnCmDBCAaBbAcBdAeBfAgBhAiBkA.

So, after having produced some vibrations roughly in the diameter of this composed motion *AB*, it will make the same progression in reverse by:

BiAkBgAhBeAfBcAdBaAbBDACnDmCpDoCrDqCtDsCxDuC.

After having produced some vibrations close to the other diameter of this motion, it will begin the first progression *CDxCuDtC*, etc. again; and in the same way, the progression will continue alternating to the left and right, while getting larger and shrinking now to one, now to the other diameter of this motion. In Fig. 258, these two diameters *AB* and *CD* make a right angle; but they can intersect at any angle, and if that angle is equal to 0, the motion is only vibratory. One can vary at will the size of this angle by small differences in the direction in which motion is produced. When the observer's eye is in the direction of diameter *AB*, one sees, while the progression by *AaBbAcBd* etc. shrinks toward this diameter, two images that are getting closer together. While it produces some vibrations roughly in this diameter, the wire seems immobile; while it gets farther away from this diameter by *BiAkBgAhBeAf,* etc., one sees two images that are getting farther away from each other; and while the motion gets closer to and farther from the other diameter *CD*, one sees at the limits of these movements, *C* and *D*, a half-transparent image of this sounding body. The space contained between these two images roughly resembles a very thin spider-web. When the eye is in an oblique direction, in relation to one of the diameters, or when the two diameters intersect at an oblique angle, the phenomenon can manifest in very different ways. By using, for these experiments, a very long metal wire, I occasionally produced the second or third mode of vibration, mixed with such a circular motion; so the nodes of vibrations stay immobile, and the movement of one part is always opposite to that of the other. Similarly, every other vibratory motion of such a rod or string divided into aliquot parts can be mixed with circular and various other motions.

None of these composed motions change the sound.

Part III
On Transmitted Vibrations, or the Propagation of Sound

Section 1: On the Propagation of Sound Through Air and Through Other Gaseous Fluids

182. General Notions of Sound Propagation

The object of the preceding part was to show how the characteristic vibrations of a sounding body and their relative frequencies are determined by the shape and by the other attributes of the body. But for the *transmitted vibrations*, which I will speak of here, it is necessary to look at the system of the body by which the sound is propagated, being of an indeterminate shape and size, and susceptible to vibrations in all directions and in all possible periods of time.

183. Air Is the Ordinary Conductor of Sound

The vibrations of a sounding body are transmitted through all matter, whether directly or indirectly adjacent. To hear a sound, a continuation of matter must exist between the vibrating body and the organs of hearing. The atmospheric air is the matter through which the impressions normally arrive at our ear, but all matter, liquid or solid, can perform the same function.

184. Sound Propagates in all Directions from the Center

In the propagation of sound through air or other gases, one is able to look at the body which produces the sound as the center of an infinity of sound rays or sound lines, in which the particles of the air, pushed by the vibrations of the sounding body, push another, and so on, so that the small contractions and expansions are transmitted from one particle to another. When the same are disturbed in a single direction only, for example, by the crack of a whip, that point will be

© Springer International Publishing Switzerland 2015
E.F.F. Chladni, R.T. Beyer, *Treatise on Acoustics*,
DOI 10.1007/978-3-319-20361-4_14

nevertheless a common center for sound waves in all possible directions, because the portion of the disturbed air is equally elastic in all directions and pushes all contiguous parts.

185. Sound Propagation Through Air Does Not Differ Essentially from the Vibrations of Air in a Wind Instrument

The *longitudinal vibrations of every extended line of the open air* do not differ essentially from the longitudinal vibrations of an air stream contained in an organ pipe or other wind instrument; one can therefore look at the objects that will be explained here as a continuation of Sect. 4 of the preceding part. The frequency of the vibrations of the air in a pipe does not depend on the diameter of the pipe; if, therefore, an indefinite enlargement of the walls does not change the frequency, one will be able remove the walls and allow the open air access to all sides, without making an essential change.

The open air therefore vibrates in the same period of time as the enclosed air, so that the sound is propagated by a space of open air, at the same time in which an air stream of the same length, contained in a pipe, vibrates, the mode of vibration being simplest in a pipe open at both ends (Fig. 14 and Par. 60). The points where the compression is greatest, in the open air, are the same as the nodes of vibration in the air contained in a pipe. The principal difference is that, in a pipe, the nodes always stay in the same place; while, in the open air, the places where the compression is greatest are always more and more remote from the body that produces the sound.

186. Air Does Not Make More or Fewer Vibrations Than the Sounding Body

The air through which the sound is propagated does not make any more or fewer vibrations than the body which produces the sound. When the vibrations of the sounding body cease, the disturbance of the air also stops. There is a big difference here between characteristic vibrations and transmitted vibrations. In characteristic vibrations, at the moment when the sounding body has arrived at the shape or at the natural density, it has completed only half of a vibration. It is necessary, therefore, that it continues this vibration, because the velocity of the motion is the greatest. When one vibration is finished, and the velocity equals 0, the shape or the density is not too different from the natural one, so that it can stay in that position; it is necessary, therefore, that a new vibration begin, and thus the motions should continue without cessation, if they are not met by any resistance.

But in vibrations transmitted in the open air, or in the propagation of sound, the greatest compressions and expansions of a particle of air take place in the same moment that the particle has the highest velocity. When one vibration is finished, and the velocity equals 0, the density has achieved a natural state. There is, therefore, no reason for new vibrations to occur, except when the air is again pushed by the vibrations of the sounding body, or when these transmitted vibrations approach the nature of characteristic vibrations, because of local circumstances, forming an echo or resonance.

187. Sound Waves

The vibrations of a sounding body produce alternating compressions and expansions in each sound ray, which are called *sound waves* (*pulsus sonori, undae sonorae*). One can represent these sound waves, which are expanding in all directions, as spherical layers that surround the sounding body. One finds the distance from one wave to another by dividing the space that the sound travels in a certain time by half of the vibrations that the sounding body makes in the same time. The reason it is necessary to divide this space by half of the vibrations is because the vibrations (that is to say the simple oscillations) of a sounding body go alternately forwards and backwards, in the way that every advance compresses the nearby air and every return expands it. But (according to Newton and Sauveur) if one wants to look at every vibration as being composed of an advance and a return, which does not conform to the actual way of expressing it, it is necessary to say that the space that the sound traverses must be divided by the number of vibrations. If the gravest sound of an organ pipe is propagated through the open air, the distance from one wave to the other (according to Newton in *Princ. Phil. Nat. Lib. 11,* prop. 50) is equal to twice the length of the pipe, or rather of the vibrating air stream contained in the pipe.

Sound waves are often compared with the concentric waves that form on the surface of water agitated by a foreign body. This comparison can serve for some as an idea of some sort, but it is not quite exact. The waves of water consist of transverse elevations on the surface, but sound waves consist of longitudinal compressions that are propagated in all possible directions.

188. Propagations of Different Tones (Timbre) of Sound

As the propagation of the different *timbres* of the sound (Par. 51) is completely unknown, for example, in the different articulations of the voice; in the attributes of the sounds of different musical instruments; and in the different sounding bodies where the mode of vibration, frequency, duration, and intensity of the sound are the same; the effect is very different.[1] Thus, we know nothing about the manner in which these different modifications of the same sound are propagated in the air and other media. L. Euler has proposed some ingenious conjectures on this subject, in his *Éclaircissements sur la generation et sur la propagation du son*, Sect. 13 (*Mém. de l'Acad. de Berlin,* 1765), and in his paper *De motu aëris in tubis*, Sect. 36 (in *Nov. Comment. Acad. Petrop.* vol. XVI). He presumes that these different modifications and articulations are caused by small differences between the scale of the densities of the air particles and the scale of vibrations with which each particle is displaced by a very small distance.

189. Sound Is Also Propagated Along Different Curves

Sound is not propagated exclusively in straight lines as light is, but also in all possible curved directions. Due to equal elasticity in all directions, every point of a sound wave can be regarded as a new center from which sound waves are projected in all spaces where they had not been previously. One can sense, therefore, a sound produced behind a mountain or behind a thick wall, although the sound will be weaker than if the sound had been propagated in a straight line. Sound propagation does not take place through the perturbations of the entire mass of the mountain or the wall, as some of us have presumed, but in the air in the curved lines on the secondary sound waves.

This has been noted by Mr. Lagrange in his *Nouvelle recherché sur le son*, Sect. 49 (in the *Mélanges de Philosophie et de Mathématiques de la Société de Turin*, vol. 11). According to Newton, in *Princ. Phil. Nat. Libr*, 11, prop. 41 and 42, a similar phenomenon can be observed in waves on the surface of water. If a vessel filled with water is divided into two parts by a separation in which there is a slit or gap, and concentric waves are produced by an agitation in one part, these waves will propagate into the other part in such a way that the gap becomes a new center of concentric waves.

The curved pipes of an organ and wind instruments, curved in different ways in which all the sound is the same as in the rectilinear case, show sufficiently that elastic fluid can vibrate along curved lines as well as in straight lines.

---

[1] The intensity of sound is the power transmitted per unit area.—*TDR*

190. Several Sounds Can Be Propagated at the Same Time in the Same Mass of Air

As in every elastic body, several very small motions can be propagated by the same mass and by the same particles of air, without impeding the motions of one by the other. The same can be observed in water waves. If waves are produced by agitations in two points, the concentric circles which depart from these two centers will intersect one another without the disturbing of one by the other.

191. Uniformity of Motion

Motion is always uniform in the propagation of sound, such that the distance traversed is proportional to the time. Whether sounds are strong or weak, grave or acute, they are propagated in the same manner and with the same speed.

192. Speed of Sound According to Ordinary Theory

Since Newton, many geometricians, including the distinguished Mr. Poisson (*Journal de l'École polytechnique*, vol. VII), have been concerned with the method of determining the speed of sound theoretically. The definitive result of this research is that the speed is always equal to $\sqrt{\frac{gh}{D}}$, where D is the density and gh is the elasticity of the air, which is equal to the pressure of a column of mercury in the barometer in which h is the height and g is the gravity. The calculation results in 880–915 ft/sec (slightly more than 288 m).

193. Results of Observations

The results of observations [of the speed of sound] have always exceeded those of ordinary theory. The very exact observations of Cassini, Maraldi, and la Caille, reported in the *Mém. de l'Acad. de Paris*, 1738 and 1739, have yielded a velocity of 1038–1041 ft (338 m)/sec. In the not-very-exact observations that were made by Müller at Göttingen, making use of a thirds watch, the velocity was 1040.5 ft/sec and the difference of several observations was no greater than six thirds; these were reported in the *Notices litteraires de Gottingue* (*Göttingsche gelehrte Anzeigen*), 1791, no. 159, and in *Voigt's Magazin für das neueste aus der Physik und Naturgeschichte*, vol. viii, ch. 1, p. 170. Several other observations made by Derham, Flamstead, Bianconi, Conmadine, etc. give results that are still greater.

194. Circumstances That Affect the Speed of Sound

The speed of sound propagation in air can only be changed by changing the specific elasticity, that is, the ratio of the absolute elasticity to the density. Here belongs especially the expansion of the air by heat, which increases the specific elasticity and diminishes the density, while the pressure remains the same. This is the reason why, according to the observations of Bianconi (*Comment. Bonon*, vol. II, p. 365), the speed is greater in summer than in winter. The mix of the types of gas, weighing more or less, can also produce changes in speed.

The strength, weakness, depth, or height of the sounds, and in general the nature of the primal disturbance, do not influence the speed of the propagation of the sound. On high mountains, and in general at high elevations, the speed is the same as it is in the lower air, because at more or less great height (supposing that the heat and the mixture remain the same) the absolute elasticity and the density increase or diminish to the same degree, so that the specific elasticity does not change. The direction in which the sound is produced is not influenced by the speed. One hears, for example, a cannon shot in the same interval of time, whether it is discharged toward one side or the other. The accelerations or decelerations that are made by the wind are not greater than its speed. The quality of the weather does not seem to influence the determination of the speed; the same speed is observed in foggy or rainy weather as it is in calmer weather.

195. Ways of Explaining the Differences Between Observation and Theory

The theory, from which we deduce the speed of the propagation of sound, seems to be too much in compliance with the laws of mechanics for us to abandon it. One makes, therefore, several assumptions for explaining the differences between the results of the theory and those of observations, some of which are listed below.

1. *Perhaps the air contains several solid or liquid particles which increase its gravity without changing its absolute elasticity, and which transmit the sound instantaneously; it is necessary, therefore, to look at the air as it would be without these particles being mixed in.*

If air had a mixture of solid or liquid particles, it would be less transparent, especially when it was condensed by the cold, or compressed by mechanical means; one would also find similar particles in the chemical analysis of the air, which is not at all the case. The observations also show that, in rainy weather or in thick fog, the speed of the sound stays the same.

2. *Sound is looked at ordinarily as a simple pulse, transmitted in the air, but when several follow on each other, one accelerates the other.*

L. Euler proposed this hypothesis in his *Conjectura physica circa propagationem soni et luminis*, Berol. 1750; but he himself regards it as false in his *Dissertation de la Propagation du son*, Sect. 42 (in the *Mém. de l'Acad. de Berlin*, 1759). If such an acceleration of one pulse by another took place, acute sounds would propagate more quickly than grave sounds, which is in contrast to the observations.

3. *In the theory, very small disturbances are assumed, but a very strong sound, such as those which have been observed, must advance more rapidly.*

But the theory and the observations show that the speed does not depend on the strength or on other attributes of the sound.

4. *Elasticity is always assumed to be proportional to the density, but perhaps there are some distortions in the different degrees of compression.*

Some want to claim that air is condensed a little more or less than the degree of the compression; but up until now Mariotte's Law[2] has always held constant, when the compressed air is at rest and the temperature stays the same.

Mr. Poisson (*Journal de L'École Polytechnique*, Chap. 14) and Mr. Biot (*Journal de Physique*, vol. LV, p. 173) have shown, following the ideas of Mr. de Laplace, that if thermal effects are entered into the calculation, it is done in every compression of the air and that the elasticity increases. The results of this theory are in accordance with the observations. The results of the research of Mr. Biot on the propagation of sound through vapor (in Vol. II of the *Mémoires de la Société d'Arcueil*) showed evidence of the existence of thermal effects in the propagation of sound.

It seems to me that a thermal effect, through compression, is an unknown chemical quality, to which I have attributed the difference between the theory and the observations and which, while being modified differently in different gaseous materials, makes the air, or a similar mix of oxygen gas and nitrogen gas, seem to vibrate a little faster than the two gases of which it is composed (Par. 67).

196. The Speed of Sound in Different Gases

Following the theory, the speed at which sound is propagated through different types of gas will be the inverse of the square roots of their weight, so that a lighter gas will propagate the sound faster than a gas weighing more.

One will not always have at one's disposal a rather long expanse, filled with any type of gas, to make direct observations; but the ratios of the sounds of the same organ pipe—full of, surrounded by, and inflated with different types of gas (Par. 67)—will serve to judge the speed at which the sound would be propagated by these fluids, because transmitted vibrations are not essentially different from characteristic vibrations. Therefore, if the sound is propagated in the atmospheric air, or by an artificial mix of oxygen gas and nitrogen gas, in one second of time, over a distance of 337 m, oxygen gas (whose sound was graver by a semi-tone or

---

[2] Also known as Boyle's Law or the Boyle-Mariotte Law.—*MAB*

close to a tone) will propagate it close to 310 m/sec. Nitrogen gas will propagate it a little less than in the atmospheric air. Hydrogen gas, according to whether it is more or less pure, gives more acute sounds by close to an octave or a tenth; the speed of the propagation of the sound in this gas will therefore be between 680 and 820 m/sec. In carbonic acid gas, which gives a graver sound by a major third, the speed will be 269 m/sec; and in nitrous gas it will be close to 320 m/sec.

These frequencies are not completely in accordance with the theory; it appears, therefore, that they are dependent not only on the ratio of the elasticity to the gravity, but also on the chemical attributes of every gas. I have nevertheless admitted that it would be useful to repeat the experiments, which I made with Jacquin of Vienna, more carefully this time, using a monochord divided in decimals for the comparison of the sounds.

Mr. Perolle, who has published many very interesting experiments on the propagation of sound in different materials, has disputed these experiments, believing the results contrary to his. But it is only a misunderstanding, caused by a very short abstract that was given on my paper. My experiments concern the *speed* of the propagation of sound by different gaseous materials, speed that, following the theory, has to be the inverse of their gravities, and this is what the experiment notes, with a few differences caused by chemical attributes.

The experiments of Mr. Perolle concern the *intensity* of the sound propagated by the same materials, intensity that, according to the theory, must be greater in heavier gases than in lighter gases, and this observation (which will be the question of Par. 198) also notes a few exceptions caused by chemical attributes. Therefore, it is not a disputed subject; it is even necessary that, apart from several differences, the heavier gases propagate the sound with less speed and more intensity; and the lighter gases with more speed and less intensity, when the absolute elasticity is equal to the atmospheric pressure.

197. Intensity of Sound Transmitted in the Air

The *intensity* of sound transmitted in the air depends on the following circumstances:

1. The *size* of the sounding body. The greater the surface that shakes the air with its vibrations, the greater the intensity of the sound. This is the reason why the sound of a string stretched on a resonant board is stronger than the sound of the same string stretched on a narrow piece of wood that is not in contact with other resonant bodies; and the sound of a tuner alone is weak in comparison to the same tuner when it is leaning on a table or on other solid bodies to which it can transmit its vibrations.
2. The *intensity* of the vibrations of the sounding body. The greater the oscillations on the vibrating part, the greater the displacement of the contiguous air particles, and, consequently the compression of all the others increases.

3. The *number* of vibrations. When the vibrations are in a more or less rapid mode, where each simple vibration shakes the air with the same force, an acute sound will have more of an effect on the ear than a grave sound, because of the greater number of vibrations. If one wants the intensity of the grave sound to be the same as that of the acute sound, the intensity of the simple vibrations must be the inverse of the number of vibrations given in the same time interval. The best research on this subject, and on its application to string instruments and wind instruments, is located in the work of Giordano Riccati, *Delle corde ovvero fibre elastiche, Schediasme* VI.

4. The *distance*. Following the ordinary theory, the intensity of the sound, as the intensity of every motion that leaves a common center in any direction, diminishes by the squares of the distances.

5. The *density of the air*. It is supposed that the intensity of the sound is because of the density. The sonorous vibrations of a body, produced in an empty space, will not be heard if one prevents the transmission of these motions to the solid matter in which the sounding body is suspended or on which it is placed. Zanetti has observed a weakening of the sound in an open vessel, where the air was heated, and consequently its density diminished without changing the absolute elasticity (Hawksbee, *Exper.* vol. II, p. 323). Roebuck (*Transactions of the Royal Society of Edinburgh*, vol. V, part 1, p. 34), being shut up in a cavity excavated in rock, which served as an air reservoir for the iron foundries at Devon, has observed that the intensity of the sound was considerably augmented by the air compressed by the action of the bellows. Because of the different densities of the air, a sound produced down low is heard better, at a greater height, than the same sound produced at the greater height down below.

6. The *direction* in which the air is pushed by the vibrations of the sounding body. One hears, for example, the sound of a glass or metal plate more strongly if the ear is turned in the direction of the surface. The discharge of a cannon fired directly towards the observer is distinguished by greater intensity than that of a cannon fired in the opposite direction. One will hear the words better if the speaker is turned towards this side, etc.

7. The *direction of the wind*. It appears that wind increases the compressions of the air by a modulation of the sound waves in such a way that, if the direction of the wind gets behind it, a larger quantity of sound waves touches our ear than in the opposite case.

198. Intensity of Sound Transmission Through Different Types of Gases

The *intensity* with which sound is transmitted through different types of gas depends, following Priestley (*Experiments and observations relating to the various branches of natural philosophy,* vol. III, p. 355), exclusively on the density. For his experiments, he made use of a ring bell and a hammer, set in motion with a timing device, and contained in a glass bell filled with gas, to observe the distance at which the sound ceases to be perceptible to the ear.

In hydrogen gas, the sound was not much more perceptible than in the empty space; in carbonic acid gas, it was much stronger than in the atmospheric air; and in oxygen gas, it was also stronger than in the air, to a degree which, it is believed, surpasses that of the density. Experiments made in nearly the same manner by Mr. Perolle (*Mém. de l'Acad. de Toulouse*, 1781; *Mém. de l'Acad. de Turin*, 1786–1787; *Journal de Physique,* vol. XLVIII, p. 455) gave the following results, which differ a little from those found by Priestley:

	Tone	Intensity	Distance at which the sound is no longer perceptible
Atmospheric air		....	56 ft, 6 in.
Carbonic acid gas	A little lower	More muffled	48 ft, 5 in.
Oxygen gas	Seems a little more acute	More clear	63 ft
Nitrous gas	Nearly as in oxygen gas	....	
Hydrogen gas	Not distinguished very well, and resembling rather a very faint noise	Without consistency and without strength	11 ft

If the sound of a small hand bell was a little graver in carbonic acid gas, I attribute it to the delay that the vibrations of a sounding body, as also the oscillations of a clock, feel from the resistance of the gas that surrounds them, which is so much greater when the fluid is denser. If the tone of the same sounding body seemed a little more acute in the oxygen gas, this might be due to a type of illusion in the greatest intensity of the sound. Otherwise it would be as difficult to explain as the difference of the results that Priestley and Perolle obtained as to the intensity of the sound in the carbonic acid gas. The other results show that the intensity of the propagation of the sound is greater when the fluid is denser, just as the theory claims.

Nitrogen gas was not examined by either Priestley or Perolle; however, it can be assumed that the sound propagated by this gas will be a bit weaker than if it had been propagated by atmospheric air.

In my experiments mentioned in Pars. 67 and 196, which, like those of Priestley and Perolle, do not deal with sound (of another sounding body) *in* different gaseous materials, but the sound *of* these materials, I did not aim to examine the intensity of the propagation of sound by these materials. However, these experiments could have indirectly had such an effect, because the sounding part of the gas enclosed in the organ pipe was surrounded by the other mass of the same gas, which propagated the sound. I remember precisely enough that the sound of hydrogen gas, though much more acute, was very weak and difficult to distinguish, and that the sound of the oxygen gas, though more grave, was stronger than that of the atmospheric air.

In the sound of carbonic acid gas and nitrogen gas, I do not remember having observed anything remarkable as to the intensity. The nitrous gas seems better suited to propagating sound than in making sound itself.

One inspired observation on the effect of hydrogen gas, which yielded the sound of a spindly and reedy voice, is reported by Odier in the *Bibl. britannique*, nos. 79, 80, p. 347, and in the *Journal de Physique*, vol. xlviii.

199. Sound Propagation Through Vapors

The *vapors* of water, alcohol, and ether propagate the sound as well as air or other gaseous materials, according to the experiments of Mr. Biot, published in Vol. II of his *Mémoires de Physique et de Chimie de la Société d'Arcueil*, p. 94. These experiments establish Mr. de Laplace's idea that, in the propagation of sound by expandable gas, small compressions made by the vibrations of the sounding body cause heat to develop, which somewhat accelerates the speed of the propagation. For (according to Mr. Biot) it is proven by the experiments of Deluc, Saussure, and Dalton that the amount of water vapor, or the vapor of any other liquid, that forms in an empty space depends only upon the space's temperature and dimensions, so that if the vapor is compressed and the temperature stays the same, one part of the vapor will return to a liquid state. Therefore, each oscillation of a sounding body, contained in the same environment, will diminish the space in one direction and make it larger in the other direction. Thus there will be, on one side, a small amount of vapor that will return to a liquid state, and on the other side, a small amount of liquid will become vapor. These compressions and expansions will take place very close to the sounding body in the very small range of its vibrations, but they will not be produced beyond that. As such, the shock will not propagate in the rest of the fluid mass, and consequently the sound will not be transmitted. But if the sounding body, by compressing the vapor with its rapid vibrations, mechanically gives off a certain amount of heat, the small compressed portion will not pass into a liquid state; the sound will be able to be transmitted by the entire gaseous mass, just as well as it can be in permanent gases. So then, if it happens like that, it necessarily follows that in the small vibrations of vapors and gases, there is a release of heat such that it must be considered in theory as causing changes in elasticity. But trying to verify these changes by the application of a thermometer would be in vain, because this instrument is no more affected by these successive and momentary variations in heat than a barometer is by the momentary variations in elastic force that occur in the production of sound.

200. Distances at Which Sound Can Be Perceived

The *distance* at which a sound transmitted in the air is still perceptible to the ear depends entirely upon the intensity, and consequently upon everything that serves to determine it. The direction of the wind contributes a lot to augmenting or diminishing the distance at which sound can still be heard. So do certain local circumstances, for example, the directions of mountains.

There are examples where sounds have been heard from very great distances. For example, in a siege at Genoa, the cannon shots were heard from a distance of 90 miles from Italy (*Philosoph. Transact.*, no. 113). In the siege of Mannheim in 1795, they were heard at the other end of Swabia, at Nördlingen and Wallerstein. In the battle of Jena, they were heard between Wittenberg and Treuenbrietzen; I myself heard cannon shots in Wittenberg, at a distance of 17 miles from Germany (15 of which make a degree of the equator), not so much through the air as by feeling the shaking of solid bodies by putting my head against a wall. When a fiery meteor exploded, some heard the explosion 10 minutes later, from which one can judge the distance and the necessary intensity for this effect. Among musical sounds, there is none that can be heard farther away than horn music in Russia, quite perfect in its genre. It has been heard at times at a distance of 5–7 wersts, or roughly 1½ leagues [5.18 miles].

A rather well-known means of hearing a sound at a greater distance is to put one's ear to the ground. In this way, Franklin heard the sound of two stones thrown one on top of the other at a great distance, as distinctly as if it had happened right next to his ear (*Experiments and Observations,* London, 1774, p. 445). The earth's surface serves as a resonant table by transmitting the shocks that it receives.

201. Megaphones[3]

The human voice can be transmitted a great distance by a *megaphone.* To obtain this effect, it is necessary that the lateral compressions of the air, which without this would be dispersed in every direction, press against the walls of this pipe in such a way that they take, as much as possible, a direction parallel to the axis and reinforce the sound going out towards this side. A cylindrical or prismatic pipe, and in general a pipe in which the diameter is the same everywhere, could not be used for this effect; the sound waves that go out of the mouth in C (Fig. 259), in the directions F and G, and that would be dispersed, remain in the pipe, and after being pressed several times against the walls, they are dispersed while going out in the directions M and N and in all other directions. Consequently, the only effect that one can obtain by such a pipe consists in hearing the sound produced in C, at the other end in L, without the least weakening or even a little more sonorous than if it was produced in that place. Through such a pipe, two people, placed at opposite ends, could

---

[3] In French, literally "voice carrier."—*RRB*

transmit and receive words at a considerably great distance. It is therefore more agreeable to name such a pipe *communication pipe* rather than *megaphone*. This quality of a pipe with a diameter that is equal everywhere, to transmit sound with the same force at considerable distance from one end to the other, was observed by several people in the *Aqua Claudia* [aqueduct] in Rome and by Mr. Biot (*Mém. de la Soc. d'Arcueil*, vol. ii) in an aqueduct in Paris where even the weakest sounds are sustained at a distance of 951 m or 488 toises,[4] and the deepest voice, the same as when one is speaking to the ear, is heard in such a way that all words are perfectly distinct and a conversation can be followed. But to transmit a sound through the open air at a great distance, the pipe must get bigger at the other end.

According to Lambert (*Mém. de l'Acad. de Berlin,* 1763), the most suitable shape is a truncated cone, because if one wants to apply the catoptric principles, sound waves are reflected by the walls such that, after one or more refractions, they become parallel to the axis (Fig. 260) or at least only slightly divergent. If we express the cone's angle with p and the first angle of incidence by q, the angles of incidence are $q, q\text{-}p, q\text{-}2p, q\text{-}3p$, etc., until the series becomes negative; so the angle with the axis diminishes with each reverberation. All shapes that, by getting larger, turn their convexity towards the axis must be rejected, according to Lambert, because they spread the sound over an entire hemisphere. These types of shapes are good for musical instruments because it is important to spread the sound out as evenly as possible; but megaphones are designed to direct the sound towards the place where it is intended to be heard. Thus, the curvature must be such that it turns the concavity towards the axis, without actually becoming parallel to the axis, or shrinking after having gotten bigger until a certain point. For if the surface becomes parallel to the axis, it begins to produce the effect of a cylinder, and if it converges toward the axis, it will have the effect of an upside-down cone. A parabolic megaphone, whose mouthpiece must be in the focus, will have less of an effect than a conical megaphone of the same length.

Mr. Hassenfratz (*Journ. de Physique*, vol. LVI, p. 18) conducted many experiments by placing a watch in a megaphone and measuring the distance at which an ear stops hearing the ticking of this watch. If, as Lambert says, it is necessary to reject all shapes that, by getting bigger, turn their convexity toward the axis, the earpiece that is usually placed at the end of a megaphone would be useless. However, Mr. Hassenfratz observed that between two equal megaphones, the one that has an earpiece made it so that the ticking of the watch could be heard at a distance roughly double that of the other, which didn't have an earpiece. When the tinplate megaphone was doubled by wool fabric, the effect did not change.

Some authors have claimed that a megaphone has to be made of a very elastic material to reinforce the sound made by its vibrations; others claim that the body must not be elastic in order to avoid the confusion of sounds that the pipe's vibrations can cause. But it seems that this criterion doesn't matter; the pipe's own resonance contributes very little to augmenting the sound because it would also

---

[4] A pre-metric system unit in France for measuring length, area, and volume.—*RRB*

propagate toward the exterior; but it could not cause too unfavorable an effect because all solid bodies can also propagate articulated sounds. It would not even matter if the interior surface were polished or not. The intensity of the transmission of the sound will depend only upon the interior shape, but the difference in material could vary the tone.

The way of explaining a megaphone's effects following the laws of catoptrics, like the refraction of light in a mirror, while yielding valid results, does not seem to exactly conform to nature, because:

1. The refraction of light depends on each point on the surface, but the action of sound depends on the general shape of the surfaces it comes in contact with, and the effect is not changed by small imperfections in these surfaces.
2. Light only spreads in a straight line, but sound, due to new centers of sound waves, spreads in all possible directions.

It seems therefore that these changes in a sound's direction are more similar to the movement of waves on the surface of water, which, after encountering an obstacle, make secondary waves that eventually spread over the entire surface of the water, and whose center is at the same distance from one side of the obstacle as the center of the primary waves is on the other side.

We communally consider Sir Samuel Morland to be the inventor of the megaphone; his instrument was a rather large type of trumpet, made of glass or copper. He wrote about it in a paper that appeared in London, and there is an excerpt of it in the *Philos. Transact.*, no. 79, p. 3056. Athan. Kircher claims to have made similar instruments before Morland, but everything that he and others said about these instruments before Morland has more to do with ear trumpets than megaphones. Cassegrain (*Journ. des Savans*, vol. III) gave them a hyperbolic shape and claimed to have had more success than Morland. In Germany, Sturm (*Colleg. curiosum*, Nuremberg, 1701, p. 2) and J. M. Hase (*de tubis stentoreis*, Lips. 1719) made many attempts at different shapes of megaphones.

Lambert published a lot of research in his paper *Sur quelques instrumens acoustique* (*Mém. de l'Acad. de Berlin*, 1763), but it is impossible to agree with him when he wants to attribute the same movements to sound as he does to light (Sects. 9–12), and when he aims to explain the sounds of trumpets by the vibrations of metal reflected by air (Sects. 13–15). At that time, he had not yet done research on wind instruments that was as precise as what he published in the *Mém. de Berlin*, 1775. Lambert's dissertation was translated into German and published by Mr. Huth, Professor of Mathematics at Frankfurt-on-Oder (in Berlin 1796), with interesting additions. The first addition has to do with Alexander the Great's horn or megaphone, mentioned in Kircher, *Ars magna lucis et umbrae* and in his *Phonurgia*, adapted from a work attributed to Aristotle, *Aristotelis secreta ad Alexandrum magnum*. The second addition contains experiments on an elliptical megaphone, by which

(continued)

the sound was strengthened very little but was more resonant; it worked better as an ear trumpet. In the third addition, Mr. Huth recommends the megaphone as a transmitter of news at great distances through the use of intermediary stations; such a *telephone* could be useful, especially when, due to fog, etc., it is not possible to use telegraphs.

Some remarks on the megaphone can also be found in Euler's dissertation, *de motu aëris in tubis* (*Nov. Acad. Ac. Petrop.*, vol. XVI) and in Matthew Young's *Enquiry into the principal phenomena of sounds and musical strings* (Dublin, 1784), p. 1, Sect. II.

202. Ear Trumpets

An *ear trumpet* is, so to speak, a reverse megaphone, arranged so that all the action of the sound that is made on a larger surface concentrates in the auditory canal of people who are hard of hearing. Lambert (in the paper cited, Sect. 69) recommends the parabolic figure as the more advantageous, but it is necessary that the parabola is truncated near the focal point, and that at this point a small tube is adapted for transmitting to the auditory canal. One could still obtain the same effect in giving these instruments a conical figure, but it is necessary for the cone to be truncated so that the sound does not turn back before reaching the ear. Mr. Huth observed that an elliptical megaphone worked well as an ear trumpet. In practice, other shapes are used as well; for example, shapes that turn their convexity towards the axis are used more or less successfully. Sometimes this instrument is also given a winding shape for more convenience. In *Traité des sens,* by le Cat, p. 292 and in *vol. II* of *Machines et Inventions approuvées par l'Académie de Paris,* p. 109, etc., many ideas and representations of ear trumpets can be found.

203. Speaking Chambers[5]

These changes in the direction of sound through the use of ear trumpets resemble that which is observed in some rooms or chambers, called *speaking chambers*, in which a weak sound, produced in a certain spot, is heard in another very far away spot, while elsewhere it is not perceptible to the ear at a much lesser distance. The most remarkable examples, reported by several authors, are the dome of St. Paul's

---

[5] Also known as *whispering galleries.—MAB*

Cathedral in London, where the ticking of a watch can be heard from one side to the other; the gallery in Gloucester, where two people speaking very low can be heard from one end to the other, at a distance of 25 toises [approx. 150 ft]; a hall in the Paris Observatory; the Agrigento Cathedral in Sicily, where a very low voice can be heard distinctly from one end of the church to the other. Abbé Actis described the phenomena that he observed in this church in the *Mém. de l'Acad. de Turin*, 1788–1789; he also mentioned the cave known as *ear of Dionysius* (*grotta della favella*) in the ancient Latomies[6] of Syracuse. In the past, a person who stood in the center of the spiral of this cave could hear people who were standing in the convergent spirals; but this monument seems to have changed a lot due to new holes that were made below, and to those that were blocked above; and yet a small noise still multiplies infinitely. The tearing of a piece of paper can be heard very distinctly from one end of the cave to the other, though the length is 47 ft 7 in..

If one wanted to deliberately construct a hall for this effect, one could give the walls and ceiling the shape of an elongated ellipsoid, by which the sound coming out of one focus would be concentrated in the other. A shape resembling two cones, or two pyramids joined at the base, would produce the same effect, in such a way that two people, placed at narrow opposite ends, could hear each other from one end to the other with the lowest voice, while nothing could be heard in the other areas of the hall.

204. General Explanation of an Echo

When the same sound is heard more than one time, it is called an *echo*. In all cases, the reaction of the air compressed by the first sound surpasses the degree of the natural density; therefore the retrograde compression can be heard as a second sound similar to the first. When these repetitions happen one after the other, too quickly to distinguish the intervals of time, they are called a *resonance*. It is generally presumed that, at most, eight or nine different sounds can be distinguished in a second of time; therefore, in order for a repetition of the same sound to be heard, not as a resonance but as a genuine echo, it is necessary that the repetition follow the sound at least 1/9 of a second later.

Ordinarily, the reactions of a sound are explained, following the laws of catoptrics, like reflections of light on a mirror (Par. 200); but Lagrange showed, in his *Recherches sur la nature et la propagation du son*, Sect. I, ch. 2 (*Miscell. Taurinens.*, vol. I), that a true *cataphonic* or *catacoustic*, comparable to the catoptric, does not exist, as d'Alembert already noted in the *Encyclopedia*. After him, L. Euler outlined the theory of the echo, in the *Mém. de l'Acad. de Berlin*, 1765, and in his Dissertation *de motu aëris in tubis* (*Nov. Comment. Acad. Petrop.*, vol. XVI), ch. 2. Mr. Poisson produced some scholarly research on the reflection of sound by

---

[6] Quarries.—*MAB*

an indefinite plane, an ellipsoid, a paraboloid, and a circular hyperboloid of revolution, in Vol. VII of the *Journal de l'École Polytechnique*, p. 350. When an echo is formed by the reaction of air that presses against an obstacle, the retrograde compression will follow the normal laws of reverberation; the explanation by catoptric principles does not therefore give false results. But there are still other cases where an echo is formed when a long enough air stream, isolated toward the sides, does not press against an obstacle, being terminated by the open air; from which it follows that the echo is not produced by reverberations, but by other circumstances.

205. Different Cases in Which an Echo Is Formed

An echo is produced when the compression of air particles, and the speed with which each particle is displaced in a very small space, do not make an equal progression, as in the normal propagation of sound (Par. 186), but are interrupted by some sort of obstacle. Since the vibrations of enclosed air and open air follow the same laws (Par. 185), the repetitions of the same sound, caused by different progressions of speed and compression, will be presented here as reported by Euler, by considering (under varying circumstances) the movements of a mass of air enclosed in a pipe after striking it. It will be the same whether one imagines such a pipe to be straight or curved, wide or narrow, or if the mass of air is isolated from the other atmospheric air in any sort of way. When the pipe is immensely long on both sides, the condensations and the speeds always follow an equal progression, and at the same moment when the speed of the very small motion of each molecule becomes equal to 0, its natural density is re-established. There is therefore no reason that this portion of the air can continue the motion; one will therefore only hear a simple sound, after P/K seconds, if P represents the distance between the ear and the place where the sound is produced, and K represents the space that sound traverses in one second. This is the normal propagation of sound, where the air does not make more vibrations than the sounding body (Par. 186). But if the pipe tapers towards one end or towards both, the compressions and the speeds of the air particles don't follow an equal progression, and consequently the compression and the speed become equal to 0 at the same time, or until other obstacles stop the motion. A tapered end of a pipe can be either open or closed. If it is open, the air, because of its interaction with the open air, will always have the natural density, regardless of speed. If it is closed, the speed of the air in that place will always be equal to 0, regardless of compression. The different cases in which one hears repetitions of a simple sound are:

1. In a pipe (Fig. 261) that is tapered on one end Bb and is open and the other end a extends to infinity, a sound that is agitated in L will only be heard as a simple sound when it is near to the tapered open end Bb, after the time of Lb/K seconds. Toward L, and even in L, the sound is heard twice, in such a way that the resonance that is formed changes into a more pronounced echo as the listener gets farther from

Bb. In each place *a* behind *L*, the original sound is heard after the time of aL/K seconds. If *Lb* equals 338 m, there will be a two second echo. So this is an example of an echo that cannot be explained by reflection.

This case could not exist for an experiment, except if, in the place of an infinite extension, one imagines an extension too long for the sound to be able to have an effect all the way to the other end; for example, in a very long gallery, where the farther one goes from the open end, the more pronounced one hears the repetition of sound.

2. When the pipe is tapered and closed in *Bb* (Fig. 262), and stretched to infinity in the other direction, the same phenomenon will also happen. If the point *L*, where the sound is produced, is slightly farther away from the closed end *b*, one hears a resonance, and the greater the distance *Lb*, the more one hears a pronounced echo. For each point *a* and in *L* itself, the echo will follow the original sound after the time of $2BL/K$ seconds. We can see therefore that the results are the same, when the *Bb* end is closed, as if it were open. Here one could say that the echo comes from reflection; but since the same echo is formed when the end is open, although no reflection could take place there, we can see that the formation of an echo cannot really be attributed to reflection.

This second case is that which most likely occurs in experiments. This case incorporates all the cases where the open air (whose vibrations follow the same laws as those of enclosed air) pushes against a wall, a thick forest, or a rocky slope and where it is necessary to look at this mass of air as tapered on one side and indefinite on the other. The polish and the inequalities of the surfaces don't matter, since often the best echoes are found in mountainous regions and forests where there isn't a single even surface; but the general shape of the objects against which the air is pushing must be suitable to produce this effect. Some aeronauts[7] have observed that at a sufficient height the earth always produces an echo that can help judge how high the listener is.

3. When a pipe *AB/ab* (Fig. 263) tapers and is open on both ends, each sound produced in any place *L* causes a multiplied echo, in which every fourth pulse is the same. If the ear is at end *A*, it will hear the principal sound after a time of AL/K seconds, then after a time of $2BL/K$ it will hear the first echo, which will be followed by the second after a time of $2AL/K$, then the third after a time of $2BL/K$, then the fourth after a time of $2AL/K$ seconds, and so on. The principal sound will therefore be repeated an infinite number of times by the echoes, which will follow each other alternatively in intervals equal to $2BL/K$ and $2AL/K$ seconds. If the sound is produced in *A* itself, the number of echoes is reduced by half, and the intervals of time between them will be equal, and equal to $2BA/K$ seconds, so if the length of the pipe were 169 m, the echoes would follow every second. If the first sound is produced in *L*, and if the ear is in the same place, the first echo will follow the original sound after a time of $2AL/K$, the second after a time of $2BL/K$, the third

---

[7] The word *aéronaute* was coined in France in the late eighteenth century to describe the pilot (and, by association, the passengers) in a hot air balloon.—CBH

after a time of $2AB/K$, which, since it is produced by two equal agitations that are similar to the principal sound, will be stronger and more distinct. This one will be followed in the same order by echoes after the times $2AL/K$, $2BL/K$, $2AB/K$, and so on. If the point L is in the middle of the pipe, all of the echoes will follow each other in intervals equal to AB/K seconds. If the ear is in another place P, the number of echoes will be even greater, and the intervals between them will be more unequal. The four echoes would follow each other in equal intervals of time only in the case where L would be in the middle of the pipe and P in the middle between L and B.

The number of repetitions would be infinite if there were no resistance. This multiplied echo cannot be explained by the catoptric principles.

This third case, where a pipe is tapered and open on both ends, is applicable to long, vaulted galleries, open at both ends. Even on narrow paths, a resonance has sometimes been noticed that would be a true echo if the path were long enough.

Mr. Biot (*Mém de la Société d'Arcueil,* vol. ii, p. 403), in his experiments on the propagation of sound in a Parisian aqueduct, 951 m or 488 toises long, noted that by speaking in the pipe he heard his own voice, repeated by echoes, up to six times. The intervals between these echoes were all equal to one another and almost exactly half a second. The last one came back a little bit after three seconds, which is to say in a time equal to that which the sound used to transmit itself to the other end; but the person who was at that end, and who was being spoken to, only heard one sound. The half-second intervals of time that Mr. Biot observed are much less than those that would result in theory ($5\frac{1}{3}$ sec); it seems that we must attribute this to the nodes of vibration that were formed in the mass of air contained in the pipe, which happens in all pipes whose diameters are small in comparison to their lengths.

4. In a pipe tapered and closed at both ends, the echoes follow each other in the same intervals of time as in the preceding case, where both ends were open.

This case applies to long galleries with both ends closed, for example in mines, where I noticed a resonance that was unpleasant to the ear. It also encompasses multiplied echoes, which are occasionally found between two walls that are rather far apart from each other, or between two rocky slopes.

5. When the pipe is tapered at both ends, with one end open and the other closed, there is a multiplied echo, in which every eighth pulse or, if the sound is produced at the closed end and stretches toward the open end, every fourth pulse is the same. This fifth case can take place in rather long vaulted galleries, open at one end and closed at the other.

206. Remarkable Examples of Echoes

Some examples of particularly remarkable echoes are mentioned by Athan. Kircher in his *Phonurgia,* and by several other authors, for example, the one at Genetay, two leagues from Rouen, which change the voice in different ways; the one by Koblenz on the bank of the Rhine, which repeats a word 17 times; the one by the Chateau

Simonetta, caused by two parallel walls, which repeats a word up to 40 times. Gassendi mentions an echo near the tomb of Metella, which repeated the first verse of the *Aeneid* eight times. In the *Mém. de l'Acad. de Paris* 1710, there are notes about an echo caused by two opposite towers, which multiply the sound 12 or 13 times. Another, not far from Milan, which multiplies it even more than that, is mentioned in the *Philos. Transact.* 480, No. 8. It was claimed that the one near Rosneath in Scotland repeats a melody three times, each time in a deeper tone, which I find to be unlikely. At Muyden, not far from Amsterdam, I observed an echo, well-known in Holland, caused by a semi-elliptical wall; the two foci are a few steps apart from one another, in a somewhat oblique direction toward the semi-ellipse; the sound produced in one focus produces a very strong echo in the other focus. It seems to come from the earth, which I attribute to a slight tilt of the wall inside. It can be presumed that, if the wall had continued in the same way to make a full ellipse, the echo would be even stronger and more multiplied.

207–210. On the Construction of Halls That Are Favorable to Sound

207. It would be very useful to always know the best way *to build halls, so that sound could be heard distinctly throughout,* without sacrificing some other attributes or conventions, or things necessary for other goals. In most of the halls which have been successful, it appears to be a matter of chance and not an exact theory. A hall will be favorable to sound:

1. When it is well arranged to facilitate the natural propagation of the sound.
2. When the intensity of the sound is augmented by the resonance of other bodies, or by suitable reverberations.

> I have borrowed the method to explain this subject, as also some ideas, from a small paper of J. G. Rhode (*Theorie der Verbreitung des Schaller für Baukünstler*, that is to say *Theory of the propagation of sound for architects*) that came out in Berlin in 1800, and that I prefer to a lot of others.

208. It is very easy to arrange halls in a manner favorable to the natural propagation of sound, by obtaining suitable reinforcements by artificial means; but this natural propagation is not sufficient for very large halls, designed for great numbers of meetings. According to Saunders (*Treatise on theaters, including some experiments on sound,* London, 4), one can regard 70 ft as the distance at which an ordinary voice is still perceptible; a theater built according to that maxim would contain approximately 2000 people. When the space is not big, so that no one is more than 60 ft away from the speaker, the shape of the hall is almost indifferent, because the sound traverses this space too quickly for one to hear an unfavorable reverberation. A similar reverberation will be able to reinforce the sound in a single case, namely, when the surface against which the sound presses is a little distant, and consequently the reaction is completed too quickly to distinguish it from the fundamental sound. To prevent the resonance or echo that a reaction could cause, it will always be advantageous to arrange the seats in the form of an amphitheater, successively elevated, so that no surface is found in any part that is too big, against which the agitated air would be able to be pushed in the same instant. If the hall is not too high and too arched, the resonance or the echo that would be caused by the reaction of the sound from the high to low would be better avoided, and the sound will be spread more easily in this smaller space. The semi-circular or semi-oval shape that is ordinarily given to theaters is suitable to contain a lot of people at a modest distance from the place where the sound is produced; but the shape of a former theater of Athens with divergent walls (Fig. 264) could contain an assembly of even a greater number, at the same distance.

It is also necessary to avoid anything that can prevent the propagation of the sound, for example, decorations that stick out too much from the walls, etc.

An orchestra must not occupy too much space, so it is not too difficult to follow the same tempo, because of the time that it takes for the transmission of the sound from one end to the other. Similarly, it is almost impossible for all to follow the same tempo if one places two choirs far apart from each other, at the two ends of a hall.

209. Resonance of other bodies, for example, if the walls are paneled with thin boards, or if an orchestra is seated on wooden boards which transmit vibrations, is applicable for halls designed for music; but it will not make it easier to hear speech. In several theaters, the ancients used practical vases[8] between the seats of the spectators to reinforce the sound, according to Vitruve (Book 5, Ch. 5), but this method did not seem to be of any use.

210. Suitable reverberations are the better means of increasing the intensity of sound. Some architects followed principles that were directly opposed to those that it would have been necessary to follow. They imagined that sound that goes forward must make a retrograde action, which only makes a disagreeable resonance that degenerates to an echo, so much more pronounced when the reflecting surface is moved away from those that hear the sound. To obtain a useful increase in the intensity of the sound from the reverberations, it is necessary that every retrograde action is avoided, but that the sound that is scattered to the right, to the left, behind, and above, is rerouted towards the public by a suitable shape of the walls, and that it can be heard everywhere almost in the same instant as the principal sound.

Of all possible shapes in which a hall can be designed to hear sound, the ellipse is the most advantageous.[9] The principal attribute of an ellipse is that all of the rays emanating from one focal point are reunited at the other focal point. If it were possible to concentrate an orchestra at one of the focal points, and the whole public

---

[8] Sounding vases.—*MAB*

[9] Present day architectural acousticians do not consider elliptical halls to be the "most advantageous" shape.—*TDR*

at the other, the secondary sound would be very strong, but of no use: if the hall is small, one does not need a reinforcement of the sound, and if the hall is large, one will hear the secondary sound later than the fundamental sound, unless the ellipse is very elongated. When the sound is produced and heard outside the focal points, the resonance or the echo will be shown in many different ways. In Berlin, the theater and the concert hall, regular ellipses, will serve to observe the effects of the elliptical shape. A round shape would not be suitable because of the multiple reverberations, as, for example, in the dome of St. Paul's Church in London and in the Rotunda[10] in Rome. I have also observed a very prolonged resonance in a half-round hall.

When it is a matter of hearing a speaker distinctly everywhere, or a singer, or an instrument, a parabolic shape for the walls and the ceiling would be highly suitable[11]; one could make the two branches of the parabola pass into upright parallel lines (Fig. 265). The sound should be produced at the focal point of the parabola, marked in the figure by a dot; and all of the sound that does not directly reach the public would reach them reverberated in directions parallel to the axis.

In general, the intensity of the sound of an orchestra or of an organ will be greatly increased when placed under a deep narrow vault, which is not very wide. The same effect would also be obtained by giving the walls and the ceiling a conical or pyramid-like shape, that could be extended into parallel walls (Fig. 266); the speaker or the orchestra should be seated in the narrow section of the hall, at the point where it would be truncated or rounded; the reverberations of the sound would resemble those of a megaphone. Whatever the form, a successive elevation of the seats will always be favorable to the sound.

It appears to me that music is very effective in a round hall where the ceiling is vaulted at a sufficient height; the orchestra should be seated at the very top, or in the center of the dome. The effect would be the same if the ceiling were conical, or if the shape of the hall were a square or a polygon and the ceiling were pyramid-shaped. The sound coming from the top, and reverberating almost as in a megaphone, would be heard everywhere very distinctly without the least echo and without any prolonged reverberations.

I have observed a surprising effect of music at Ludwigslust in the church of the court of the Duke of Mecklenburg-Schwerin. The Church has a single nave; all the way at the end, where the altar is found, there is a tableau that represents the appearance of the angels that announced the birth of Jesus Christ to the shepherds. Between the boards on top that form the clouds is seated an orchestra that does not see the public, and which is not seen; all the sound is poured out from the top and only reaches the public by reverberations from the ceiling. The sound is beautiful and distinct, and before knowing of the construction of the Church, it is difficult to guess where it comes from; one looks for the orchestra without finding it.

---

[10] Pantheon.—*MAB*

[11] Present day architectural acousticians would probably disagree with this statement as well.—*TDR*

Rhode, in his *Traité* cited in Par. 207, noted that most theaters are not very favorable to sound, because the laws of the propagation of sound are neglected in the use of pipes with parallel sides and megaphones. Box seats on the sides, near the stage, are very unfavorable because they absorb too much sound. It would produce a better effect if there were only straight or parallel walls on the sides, as in Fig. 267, or diverging walls, as in Fig. 264, without protruding decorations. The ceiling should not be too high; it can be parallel to the orchestra and gradually rise toward the part of the hall farthest from the stage. The opposite end of the theater can form the arc of a circle. The well-known theater in Parma can serve as an example. For even larger theaters, divergent walls would be preferable. The amphitheatrical seating arrangement would not prevent the construction of some separate special seating, by interrupting the series of seats by a modest space. Rhode also noted that the standard arrangement of the wings [backstage] is unfavorable to the propagation of sound, because they absorb all the sound that spreads toward the sides. According to him, the old triangular turning machines, whose three surfaces can be painted or covered in painted netting, would be more favorable for sound, when they are turned in such a way that they form a wall that reflects sound forward, at least toward the proscenium.

Since an air current, going in the same direction as the sound, greatly intensifies the sound, some have wanted to use that method in a theater; but experience demonstrated that the inconveniences surpassed the advantages that could be gained.

It seems that ancient theaters, where the seats were successively elevated as they got farther from the stage, were more suitable for facilitating the natural propagation of sound than reinforcing it by artificial means. In the rest of these theaters, for example in the Circus of Murviedro in Spain (in the ancient city of Sagunto), according to Mr. Biot, and also as in the Arena of Nîmes, and in the amphitheatre in Hadrian's Villa in Tivoli, what is said in the arena can be heard very well in the more elevated areas. These effects can be attributed in part to reflections of sound by the earth, and to the fact that the propagation of sound up high is facilitated by the action of the denser air on slightly less dense air.

211. Works and Dissertations that Contain Research on Sound Propagated in the Air

The following works and dissertations serve to further inform the research that has been done on the theory of the propagation of sound in air:

H. Newton, *Principia Philosophiae naturalis mathematica*, lib. I, sect. VIII, *de motu per fluida propagato*.

Research on the propagation of sound, by L. Euler, with two continuations in the *Mém. de l'Acad. de Berlin*, 1753.

Clarifications and details on the generation and the propagation of sound and on the formation of an echo, by L. Euler, in the *Mém. de l'Acad. de Berlin*, 1765.

L. Euler, *de propagatione pulsuum per medium elasticum*, in *Nov. Comment. Ac. Petrop.*, vol. I.

L. Euler, *de motu aëris in tubis*, in *Nov. Comment. Acad. Petrop.* vol. XVI.

Research on the nature and the propagation of sound, by Lagrange, in the *Melanges de Philosophie et de Mathematiques de la Société de Turin*, vols. I and II.

On the manner of rectifying the two points of the principles of Newton, relative to the propagation of sound, and the motion of waves, by Lagrange, in the *Mém. de L'Acad. de Berlin*, 1786.

Giordano Riccati, *Delle corde ovvero fibre elastiche*, Bologna, 1767.

Enquiry into the principal phenomena of sounds and musical strings, by Matthew Young, Dublin, 1784.

On the speed of sound, by Lambert, in the *Mém. de l'Acad. de Berlin*, 1768.

Observations on the theory and on the principals of the motion of fluids, by J. Trembley, in the *Mém. de l'Acad. de Berlin*, 1801.

Treatise on equilibrium and the motion of fluids, by d'Alembert, book 2, chap. IV, and *Opuscul.* vol. V.

On the theory of sound, by Mr. Poisson, in the *Journal de l'École Polytechnique*, vol. VII.

Section 2: On the Propagation of Sound Through Liquids and Solids

212. All Possible Materials Propagate Sound

The vibrations of a sounding body are transmitted to all contiguous material, directly or indirectly. Air is the ordinary conductor of sound and the most proper medium for transmitting the sensation of sound to the exterior organs of hearing to man and to all types of land animals. But all liquid and solid materials also propagate the sound with a great deal of intensity; the same with all of the modifications to the sound propagated by these materials.

213. Propagation of Sound in Water

The propagation of sound in *water* may be concluded from the fact that aquatic animals also possess organs of hearing; it is noted also in the experiments. When under water, one can hear sounds that are produced in the air, but one hears more strongly the sounds that are produced under water. (*Journal des Savants,* 1678, p. 178; Hawksbee, *Philos. Transact.* No. 321; Arderon, *Philosoph. Transact.* No. 486; Nollet, *Mém. de l'Acad. de Paris*, 1743 and *Leçons de Physique expérimentale*, vol. III, p. 417; Musschenbroek, *Introd. ad Philos. nat.*, vol. II, Sect. 226; Monro, *Physiology of fishes, etc.*, Ch. IX).

Sound produced in the water is also heard in the air. The air does not contribute at all to the propagation of sound in water; if the air contained in water is carefully separated, the propagation is the same, according to the experiments of Nollet and Musschenbroek. But, unlike air, water is not compressible when an enormous force is applied, except to a very small degree, according to the experiments of Canton, Abich, Zimmermann, and Herbert. One will not be able to apply the theory of the propagation of sound in air, to determine the way in which one particle of water transmits the pulse to another, which is not consistent in compression and

© Springer International Publishing Switzerland 2015
E.F.F. Chladni, R.T. Beyer, *Treatise on Acoustics*,
DOI 10.1007/978-3-319-20361-4_15

expansion. These differences between liquids and expandable gases are also shown in that liquids never make the same sound vibrations as the air or other gaseous material contained in a pipe.

214. The Resistance of Water Delays the Vibrations of a Sounding Body

When a bell or a sounding vessel is full of water, or when a sounding body is plunged into water, the sound is graver than that produced in the air, because of the delay of the vibrations by the resistance of the water as a denser fluid. This delay increases when the vessel is filled with water, or when the sounding body is plunged more deeply under water. At an even greater depth, sound vibrations cease, and only an imperceptible clacking is produced. Some other liquids, for example, oil, milk, foaming champagne, etc., resist even more sound vibrations than water.

215. The Velocity of Sound Through Liquid Matter is Unknown

The velocity with which sound is propagated in water, or in other liquids, is completely unknown. One can nevertheless presume that it will not be the same at different depths, because the density does not increase due to the pressure, as in the expandable fluids. It would be difficult to do experiments on this subject.

216. The Intensity of Sound Propagated Through Water and Through Other Liquids

The intensity of the propagation of sound through water, when it is produced in the water, greatly surpasses that of the propagation of sound in the air. Nollet observed that the effect of two rocks striking against each other was almost unbearable. A sound produced in the air is also heard under water, but it is weaker, because of the lesser action of a less dense fluid[1] on a denser fluid.

Mr. Perolle did many experiments on the intensity of sound in different liquid materials, which he published in the *Mém. de l'Acad. de Turin*, 1790–1791. He used a watch hanging by a thread in a vessel filled with the liquid material to determine the distance at which one can still hear it ticking. In the air, this distance was 8 ft, in

---

[1] Gas, for example.—*MAB*

water 20 ft, in olive oil 16 ft, in oil of turpentine 14 ft, and in ethanol 12 ft. When repeating these experiments, he did not always obtain the same results. We cannot demand in these experiments the same exactitude that would be required if there were a continuation of the same material between the sounding body and the ear; yet we can see that these fluids vibrate with more strength than air; even the vessel and the table it was placed on were noticeably vibrating; the surface of the water stayed still. Each fluid is distinguished by a different tone.

Mr. d'Arnim (*Annal. de Gilbert*, vol. iv, ch. 1, p. 113) notes that the intensity of sounds must be due to the specific gravities of fluids, if the other factors remain the same, and that the results that Mr. Perolle obtained do not differ very much in these specific gravities.

> The surface of water stays still because the motion of each particle happens only in an extremely small space, such that it is impossible or almost impossible to perceive. The motions of the surface of the water, represented in Figs. 252 and 257, do not apply to the object I am talking about right now, because they are caused by sounding vibrations of the vessel itself, which pushes back the contiguous/surrounding water.

217. Solid Matter Also Propagates Sound

Solid materials propagate sound very strongly, especially if their shape is favorable to vibrations; but to better perceive the sound propagated by such a material, it is useful to press it against the firm parts of the head, which can transmit the impressions to the interior organs of hearing. A simple wire made of such a material will suffice to propagate sound; for example, when two people stretch out a wire, while holding the ends between the teeth, they can hear each other by covering their ears and speaking very low. If one hangs a large silver spoon at one end of the wire and holds the other end between the teeth, it sounds, when one's ears are covered, like the sound of a large bell. With the ear held to one end of a long beam, one distinctly hears the impact of a pin hitting the opposite end, whereas the same sound transmitted by air cannot be heard at the same distance. A rod of any length, thickness, and material transmits sound, and even words, very well if one of the ends is held against the sounding body and the other is held by the teeth or at another firm part of the head, especially when the material of the rod is somewhat elastic. The effect is almost the same if the person who is speaking holds the rod to their teeth, their throat, or even a button on their clothes, held tightly against their chest. Instead of a single rod one could also use an extension of several rods, even if they are joined at different angles. Words are heard even more distinctly if the rod is held against a metal, glass, or porcelain vessel, and if the speaker directs his voice toward the inside of the vessel. The intensity is even greater if the vessel itself is held against the teeth, or another suitable part of the head. The sounds of an instrument are heard very well, when the ears are blocked, and the end of the rod is pressed to the resonant table or against the walls of the instrument. In the same way, it would also be possible to hear the sound of a tuning fork pressed against an instrument, after its vibrations, propagated by the air, stop being perceptible to the ear. This method of hearing sounds produces a sensation almost as if the sound were coming from the rod itself. Every material changes the tone differently.

Deaf people, or those who are hard of hearing, could use this method to hear words or the sound of an instrument, if the cause of their hearing loss is located in the exterior organs; but if the interior organs are the cause, it would not be helpful to them.

This propagation of sound through all solid materials also allows the miner digging a passageway to hear the blows of the miner on the opposite side and thereby judge his own direction.

Many observations about the propagation of sounds by solid materials can be found in a dissertation by J. Jorissen, *Nova methodus, surdos reddendi audientes,* Halle, 1757; and in another by Winkler: *de Ratione audendi per dentes,* Lips. 1759; in Kircher's *Musurgia,* book 1, sec. VII, ch. 7; in *Boerhavii Praelect. in Institut. Rei medicae,* vol. IV, *de auditu, etc.* More recent studies include those of Perolle, Biot, Herhold, and Rafin.

218. Direction of Motions

One can presume that the longitudinal or transverse direction of the motion of a propagating body, when it is pushed by the vibrations of a sounding body, depends in part on the form of the propagating body, and in part on the direction in which the sounding body acts on the body that propagates the sound. The nature of the vibrations of the sounding body (if they are transverse or longitudinal) will be indifferent.

219. Velocity of Sound Through Solids

It seems to me that the *velocity of the propagation of sound in solid materials,* as long as it is made by longitudinal vibrations, can be determined by the following method. Sound is propagated by a length of open air in the same amount of time as that of a column of air of the same length, enclosed in a pipe, which makes a longitudinal vibration (Par. 185). The longitudinal vibrations of rigid bodies (Pars. 77–83) follow the same laws as air; we can thus suppose that the sound is propagated by each rigid or expandable material at the same time that that material, as a sounding body, makes a longitudinal vibration. The propagation of sound by rigid materials would thus be all the more rapid as the longitudinal sound gets more acute, supposing that the length is the same. Therefore, roughly the same correlation will exist between these velocities and the velocity of the air, as the sounds presented in Par. 82. However, the length being the same, the longitudinal sound of tin is more acute than that of air by two octaves and a major seventh. That of silver is more acute by three octaves and a tone, and that of copper by almost three octaves and a fifth. Those of iron, glass, and fir, whose vibrations are more rapid, surpass that of air by at least four octaves and a semi-tone, etc. So if there were a sufficiently long and homogeneous continuation of such a material, the speed of the propagation of sound by air would be surpassed by that of tin approximately 7.5 times, silver 9 times, leather almost 12 times, iron and glass almost 17 times, different types of wood 11–17 times, and terracotta roughly 10–12 times.

220. Experiments That Have Been Performed on This Subject

The experiments that have been done up until now note a greater speed of the propagation of sound through solid materials than through air. Mr. Wunsch, Professor at Frankfurt-on-Oder, published experiments (in his *Mémoires allemands, présentés à l'Académie de Berlin* 1793) on the propagation of sound by a very extended expanse of wooden boards. Sound is propagated much more rapidly than by air; but we cannot agree with him when he claims (like Hook in the preface of his *Micrographia*) that sound propagates itself through solid bodies instantaneously, or at least as quickly as light. Mr. Herhold and Mr. Rafin in Copenhagen performed experiments (published in Reil's *Archiv für die Physiologie*, vol. III, ch. 3, p. 178) on the propagation of sound by a cord with a length of 300 ells or 600 Danish feet. One of the ends of this twisted linen cord was tied to a wooden stake, and a silver spoon was attached near that end, so that they hit each other; the other end was pressed against the ear, or held in the teeth, while holding the cord. The sound was heard through the cord much quicker than through the air; the difference seemed to them to be almost a second, which seems to be too much for that distance. The most interesting experiments on this subject are those of Mr. Biot, which I mentioned in Pars. 201 and 205, described in vol. II of *Mémoires de la Société d'Arcueil*, p. 403. For these experiments, he used pipes in a Parisian aqueduct, made of cast iron, that altogether formed an uninterrupted length of 951 m (488 toises). In the last pipe, he placed an iron ring of the same diameter as himself, wearing at his center a bell and a hammer that could be dropped at will. So at the other end it must have been possible to hear two sounds, one transmitted by metal and one transmitted by air. He went on to verify these experiments with two demonstrations where, after a certain time, someone delivered a blow at each end. Mr. Hassenfratz (*Traité de Physique par M. Haüy,* sec. 479), having gone down into one of the quarries located underneath Paris, tasked someone to hit a hammer against a rock mass that forms the wall of one of the passageways, in the middle of the quarries. He always distinguished two sounds, one of which, transmitted by the rock, arrived earlier than the other, transmitted by air; but it also got weaker more rapidly as the observer got farther away.

221. Intensity of Sound Propagation Through Solids

The intensity of the propagation of sound through solids greatly surpasses the intensity of the propagation of sound through open air (Par. 216). The best experiments on this subject are those of Mr. Perolle, published in the *Mém. de l'Acad. de Turin*, 1791–1792, and in the *Journal de Physique,* vol. XLIX, p. 382. He used various materials, with one end touching a watch and the other touching one of the firm parts of the head; assuming the ear was not blocked, the sound was heard much better than if the sounding body had been placed in the air at a much lesser distance. The intensity of the propagation by cylinders of different types of wood seemed to decrease in the following order:

1. Fir
2. Campeche wood
3. Boxwood
4. Oak
5. Cherry
6. Chestnut

In general, the metal cylinders propagated the sound a bit less than the wood cylinders. The intensity seemed to follow this order:

1. Iron
2. Copper
3. Silver
4. Gold
5. Tin
6. Lead

The strings propagated it with less force than the solid bodies, and the intensity seemed to follow this order:

1. Catgut
2. Hair
3. Linen
4. Silk
5. Hemp
6. Wool
7. Cotton

The pieces of zinc, antimony, glass, rock salt, gypsum, and dried clay were also good conductors of sound; marble was noticeable for the small amount of force with which it transmitted the motion.

In the experiments that I performed on this subject, I observed the greatest intensity when the sound was propagated by glass rods or by thermometer or barometer tubes, and by rods of fir wood.

It seems that the intensity also depends on the shape of the body propagating the sound, if it is more or less suitable for vibrating in different ways. A rod or a blade will propagate sound much better than a shapeless mass of the same material.

Descartes has already noted (in *Epist.* p. 2, ep. 72) that the intensity of the propagation of sound by solid bodies is greater than the intensity of the propagation of sound by air, due to the greater cohesion of these bodies.

222. Reinforcement of Sound by a Resonant Board

The resonance of solid bodies is used to increase the effect of a sounding body that, without this artificial means, would have too little intensity. The sound of a string, stretched on a narrow piece of wood, with no support, would be very weak; this is the reason why one stretches the string on a thin wooden board, to increase the effect of the vibrations that the string transmits to this bigger surface. Also, the sound of a very weak tuning fork or another type of fork is greatly increased when this body is leaned on a wooden board, or on another support that is sufficiently extensive and elastic. A similar resonant body must be looked at as being of indeterminate dimensions, since it vibrates in all possible intervals of time. In every sound, reinforced by vibrations transmitted to a larger surface, the whole body resonates in motion, in such a way that it is divided into vibrating parts, alternately above and below, separated by nodal lines, almost as in the characteristic vibrations of plates, described in Sect. 7 of the preceding part. If one wants a resonant board to reinforce all sounds, especially the gravest, it must not be too small or too thick, and it must be elastic enough to easily vibrate in all modes. In observing carefully, one will find that a resonant board often reinforces some sounds more than some others; this unequal reinforcement is most likely to take place if the sounds are the same as those that the board could render if it were the sounding body. One will be able to find the places that are more or less in motion in the reinforcement of a sound, while supporting a tuner that renders the same sounds, successively, at the different points of the board and while observing the different intensities of the sound. Any wooden box will serve for these experiments. The differences in intensity will again be greater if one sets up or supports a pointed iron wire in different places, in order to produce the modes of vibration described in Pars. 69 and 70.

A resonant board will reinforce several sounds at the same time, vibrating in different modes, when one does not prevent the other (Pars. 164–176).

Maupertuis (in the *Mém. de l'Acad. de Paris*, 1724) has better explained the intensification of all the sounds made by the same board, by claiming that each sound shook only some fibers endowed with an elasticity in compliance with this sound.

223. Sound Produced by Motions in all Bodies That Can Vibrate in the Same Time Intervals

A sound that is transmitted in the air, or in solid matter, puts in motion all of the bodies that can vibrate in the same time interval. If, in the same instrument, or in different instruments that can act on one another by the air or by a continuation of other material, two strings are in unison, and one of the strings is put in motion, the other will also vibrate; because in every time interval where it can make a vibration, it is pushed again by the vibrations of the other one. The same phenomenon will take place if one of the equal sounds, or both of them, result in divisions of the string into aliquot parts. One can render visible the nature of these vibrations by putting small papers on different points of the string (Par. 37).

Another reverberating sound will also produce more or less such a resonance, because one of these bodies, after a small number of vibrations, is pushed again by a vibration of another. A strong enough sound can rather easily shake windowpanes, walls, or other objects; this happens in the case where the nature of the shaking body permits it to vibrate in the same interval of time as the body that produces the sound.

224. Vessels Can be Broken by the Voice, According to Some Authors

Some authors, such as Morhof (*Stentor hyaloclastes, sive de scypho vitreo per vocis humanae sonom rupto*, Kil., 1683) and Bartoli (*Trattato del suono e de tremori armonici*, Bologna, 1780), talked of glass vessels, thin and convex, that were broken by a very strong and sustained voice, and that this phenomenon was preceded by a very strong quivering. The sound of voice then had to be the same each time; an octave that was suited to the vessel. I have also been told of a place in the *Talmud* (*Bawa Kama*, 18) that contains discussions on the damages that can be demanded when a vessel is broken by the voice of a domestic animal; which leads one to presume that, if a similar case had never happened, one would not have conceived the idea to take up discussion on this subject.

Part IV
On the Sensation of Sound:
On the Hearing of Men and Animals

It should be noted that Chladni's exposition treatment of hearing is of value for its historical significance. Although it does contain some correct insights, it differs from modern auditory theory in a number of ways.—JPC

Section 1: Human Hearing

A. The Structure and Functions of the Hearing Organs

225. Explanations

Hearing is the sensation that vibrations produce in the ear. The impressions of the vibrations are able to be transmitted by any matter, but the air is the ordinary conductor that transmits them to the auditory nerve by the outer and inner parts of the ear.

226. Position and Parts of These Organs

The organs of hearing are situated on two sides of the head, in the portion of the *temporal bone* which, because of its hardness, is called the *petrous bone*. The parts which constitute these organs are the *external ear*, the *auditory meatus*, the *tympanic cavity*, and the *labyrinth*. The last is the destination for the sound itself, and the other parts only serve to transmit the impressions of the vibrations in the air to the cochlea.

227. The Outer Ear

The exterior of the ear (or the outer ear) is cartilage in the shape of an almost half-oval, which is designed to reinforce the sound. This section consists of a housing called the *concha*; some projections, as the exterior edge withdraws, called the *helix*; a projection almost parallel in back of the *helix*, that crosses the ear obliquely, called the *antihelix*; a projection located toward the front of the

© Springer International Publishing Switzerland 2015
E.F.F. Chladni, R.T. Beyer, *Treatise on Acoustics*,
DOI 10.1007/978-3-319-20361-4_16

auditory meatus, called the *tragus*; and another small one on the other side of the
auditory meatus, that is called the *antitragus*. The lower section of the cartilage
terminates in a fatty *lobe*. Some muscles seem to be designed to move the outer
ear, but there are very few individuals that are able to use them for this effect;
perhaps, because one loses the abilities of early childhood because of the pressure
of head coverings.

228. The Auditory Meatus

The *auditory meatus* is part cartilaginous and part bony. The housing of the concha
becomes tubular, and continues thus as far as the bony section, terminated by the
eardrum that immediately receives the vibrations of the air, for the transmission to
the inner ear. The face of the eardrum is an irregular cone. It is concave downwards
on the outside and pointed inwards. It is attached to a circle of bone that is called the
framework.

229. The Tympanic Cavity

Between the eardrum and the labyrinth, one finds the *tympanic cavity*, an irregular
cavity that is almost a half-circle, full of air, and communicating with the mouth
by a canal called the Eustachian tube. The partition that is face to face with the
eardrum presents an oblique projection called the *promontory*. Above this projection is located an opening of the labyrinth that is called the round window; it is
covered by a membrane. To the underside there is another opening of the
labyrinth, the oval window. It allows the impressions of the vibrations on the
eardrum to be transmitted to the labyrinth by an intermediate, very mobile,
mechanism composed of four bones: the *hammer*, the *anvil*, the *lenticular bone*,
and the *stirrup*.

The *hammer* is formed by a long and thin *shaft*, with the end adhered to the
eardrum; and with a *head* that is at an angle with the shaft, and which articulates
with the *anvil*. The part of the head that is slightly thinner is called the *collar*; it has
two protuberances, the *short protrusion* and the *spindly protrusion*. This can be
looked at as the stationary point of the lever. The anvil articulates on one side of the
head of the hammer, and the part opposing the two protrusions, which it uses for
support; the other is articulated by the *lenticular bone* with the *stirrup*, whose shape
resembles that of a stirrup used to mount a horse. It is at an almost right angle with
the anvil; the mobile base forms the oval window of the labyrinth, which is shaken
on the interior by its pressure. The hammer has three muscles, the stirrup a single
one, and the anvil has none at all. It seems that this device serves to allow people to

hear more perfectly, but one has heard about cases where these organs have been destroyed (in the opinion of Astley Cooper, in the *Philosoph. Transact.* 1800, vol. 1, no. 8). Sometimes the deafness has been healed (for a short time) by the perforation of the eardrum, according to, for example, Mr. Hunold of Cassel. In these cases, the sound appears to be transmitted to the labyrinth by the immediate action of the air surrounding the membrane of the round window, which Scarpa calls the *secondary membrane of the eardrum.*

230. The Labyrinth

The *labyrinth*, so called because of the complex canals, is part of the inner ear, closely enveloped by the rock of the temporal bone. It contains the auditory nerve, differently scattered membranes, and fibers in a watery gelatin. Its parts are the three *semi-circular canals*, the *vestibule*, and the *cochlea*.

The *semi-circular canals*, of which there are two, united at one end, are vertical. And the almost horizontal third contains similar membranous canals, each of which has a bulb-shaped swelling. These membranous canals leave the bony canals and are rejoined in the cavity that is called the vestibule and form a sac, called the *common sac of the vestibule*. Another smaller separate sac is called the *proper sac of the vestibule*. The *cochlea*, part bony and part membranous, rotates itself around a conical axis in a spiral that completes two and a half turns and diminishes so that the cochlea approaches a globular shape. One of its two ramps comes through the round window that is in the tympanic cavity; the other goes to the vestibule that is connected to the cavity by the oval window.

231. The Auditory Nerve

The *auditory nerve* is very short and appears to be produced from a grayish band that crosses the posterior face of the pedicle of the cerebellum. It enters the labyrinth twisted on itself, by the *internal auditory canal*, and divides into four shafts, of which two go to the bulbs of the semi-circular canals. A third, situated between the two preceding, expands into the vestibule, and the fourth, which is the continuation of the trunk, continues into the cochlea in numerous nets.

The facial nerve that enters with the auditory nerve at the far end of the same canal gives nets to the muscles of the hammer and the stirrup and forms the *tympanic rope*, a nerve net, so named because it is placed under this membrane, as a rope that crosses the membrane of a drum.

232. Ordinary Transmission of Impressions to the Inner Ear

Sound impressions are received in the following way. The vibration of the air agitated by the sounding body disturbs the eardrum. This moves the small bones in the cavity, which act on one another as levers. The base of the stirrup imprints these vibrations on the gelatinous fluid that fills the whole labyrinth, by means of the oval window. The vibrations of the eardrum also disturb the air contained in the cavity, which transmits these impulses to the round window, so that the impression is handled in two ways at the same time. The auditory nerve, the substance of which is widespread in the whole labyrinth, transmits these impressions to the brain, as the common center of all sensation.

233. Transmission of Impressions by the Solid Part of the Head

Sounds propagated by liquid or solid matter can be heard especially well when the vibrations are transmitted to the solid parts of the head, which transmit it to the auditory nerve, as discussed in Sect. 2 of Part III. The impression is stronger than that of sound propagated by the air and transmitted through the ear in the regular way. The effect (or the timbre) is made up of the original timbre and the one that is propagated through imprinting.

Mr. Perolle has conducted many experiments on hearing, experimenting on different sections, which have been published in the *Mém. de la Societé de Mé decine*. An excerpt of his paper is found in the *Journal de Phys.*, Nov. 1783. The solid parts of the head transmit the pulsations of a ticking watch better than those that are covered with a lot of flesh. The teeth, especially the incisors, are very sensitive, as are also several bones of the cranium, first vertebrae of the spine, etc. The soft parts of the mouth and the cartilaginous parts of the nose do not have any type of sensitivity. When the watch was placed in the mouth, the sound was not propagated through the Eustachian tube.

A rather strong sound, for example, the beating of a drum, is also heard faintly when the ears are blocked, by the action of the air on the solid parts of the head.

234. The Impressions Act on the Entire Labyrinth

The impressions of the vibrations are transmitted to the two windows of the labyrinth, disturbing all the mass of liquid that the labyrinth contains, as in general every pressure on a fluid is spread to the mass, in such a way that every molecule is pressed with the same force (according to Euler, *de statu aequilibrii fluidorum,* in *Comment. Acad. Petrop.* vol. xiii and d'Alembert in his *Traité de l'Équilibre et du mouvement des fluides,* Paris, 1744). One can therefore suppose that this pressure also vibrates all of the nerves that the labyrinth contains; so that it does not conform to nature to pretend that each sound only vibrates some parts. But the impressions on all of the substance can be accomplished in many different ways. If several sounds are heard at the same time, all of the motions necessary for this effect take place at the same time without one precluding the other. The same thing takes place in all of the elements of motion. It appears that the labyrinth is organized in a complex manner to facilitate many more sorts of impressions.

235. Authors Consulted

Among the older authors, Cassebohm, Valsalva, Morgagni, Duvernei, etc., one finds important research on the organs of human hearing; but to learn of the current state of knowledge on this subject, one can consult the following works:

Anton. Scarpa, *Anatomicae disquisitions de auditu et olfactu,* Pavie, 1789, the foremost work that explains the true organization of the labyrinth.

Andr. Camparetti, *Observat. Anatom. de aure interna comparata,* Patav., 1789.

Leçons d'Anatomie comparée, by G. Cuvier, vol. II.

The *Tables anatomiques* of Loder contain depictions of the organs of hearing, Tables 54, 55, 161, 162.

Abbildungen des Gehörorgans (*Depictions of the organs of hearing*) by Sömmering, Frankfurt, 1806.

C. F. L. Wildberg, *Über die Gehorwerkzeuge des Menschen* (*On the auditory organs of man*), Jena, 1795. Work used to teach physiology and pathology on these parts.

Alex. Monro, *Observations on the nervous system.* Contains microscopic observations on the structure of the nerves of the cochlea.

B. The Subject of Hearing

236. The Ear Records the Sensation of All Sufficiently Rapid and Sufficiently Intense Disturbances

All of the disturbances that are sufficiently rapid and sufficiently intense to agitate the auditory organs produce a sensation of sound. The reason why vibrations that are less rapid don't elicit this sensation appears to be that ordinarily these vibrations do not have the force necessary for this effect. To hear slow vibrations as well as rapid vibrations, it is necessary (according to Giord. Riccati, *Delle corde ovvero fibre elastiche, Schediasm.* vi) that the intensity of every vibration is simply a function of its duration, or (to express it another way) that in different sounds, the intensity of the vibrations is an inverse function of the number of vibrations occurring in the same time interval. For this reason, and because of the different organization of each individual and of each species of animal, absolute limits on the perceptibility of sound do not exist.

In the same way, it appears that a simple and rather strong shock can sometimes be heard, as in an explosion, in the crack of a whip, or in a sudden burst of the air in an empty space. One can presume, however, that a simple shock that is able to cause some irregular vibrations in solid bodies and in the air encounters different obstacles. It is perhaps for this reason that often such a simple shock is not heard in a single moment, but with some resonance, as, for example, thunder.

A progressive motion, or in general a motion that is not vibratory (Par. 1), is not heard, unless it produces vibrations in the air or in other matter. In the rapid passage of a cannon ball or a rifle bullet through the air, one hears a hissing, of which the tone, when it is distinguishable, appears dependent on the magnitude of the body. The displacement of air which is found in the direction of the motion, the burst of air behind this body, and the friction on the air to the sides excite the vibrations more or less regularly in the air, as the friction generated in solid materials. When one hits the air quickly with a cane or a stick, it also produces a hissing or buzzing. As to how much of the tone is perceptible, it seems to me to depend above all on the width of the surface that displaces the air.

237. It Records the Sensation of the Relative Frequency of the Vibrations

When one hears two or more sounds at the same time, one after the other, the ear records the sensation of the relative frequency of the vibrations (Par. 6) and of their coincidence (Par. 177). The motions act on the ear, as shapes on the eye. We do not calculate the ratios themselves; but nature calculates for us and managing to reach our sensations is the result of these ratios. An exclusive usage of consonant ratios, which because of their simplicity are pleasant on their own, would cause too much monotony. It is therefore also necessary to present dissonant ratios, which, being more complex, are not agreeable when they relate to themselves or when they pass by others which are simpler. The more or less agreeable effect of complex strong ratios is not the same for everyone; it depends on differences in organization and habit. Thus, for example, a choral fugue of Handel, which delights the connoisseurs, will be only a confused noise for those who do not know to follow the way of many voices at the same time.

Descartes (epist. 111) expresses very well the effects of consonances and dissonances: *Inter objecta sensus illud non animo gratissimum est, quod facile sensu percepitur, neque etiam, quod difficillime, sed quod non tam facile, ut naturale desiderium, quo sensus feruntur in objecta, planè impleat, neque etiam tam difficulter, ut sensus fatiget.*[1]

The measure of the more or less agreeable effects of these ratios, which Euler gave in his *Tentamen novae theoriae musicae*, conforms somewhat to the experiments.

238. Very Small Differences from the Exact Ratios of Tones Are Not Perceptible to the Ear

The ear cannot distinguish small differences in the exact ratios between the sounds. There is rather the sensation of the simpler ratio, from which the one that is heard, in effect, only differs slightly (Pars. 20 and 25). Without this illusion, there is no point to music (Par. 21).

The different absolute frequencies of the vibrations give the impression of a sound more or less grave or acute; but the same numbers are not able to be perceived by the ear, because, as with the gravest sounds, the vibrations follow one on another too quickly to distinguish (Par. 5). One is able to count 8 or 9 vibrations within about one second's time. But the gravest sounds which one is able to hear make at least 30 vibrations per second. All sounds distinguishable or perceptible by us are contained in little more than nine octaves; but we do not know if there are living beings to whom vibrations a lot slower or quicker are perceptible as distinct sounds.

239. Ordinarily, the Shape of a Sounding Body and Its Mode of Vibration Cannot Be Determined by Hearing

In most cases, the shape of a sounding body and its mode of vibration cannot be determined by hearing. The fundamental sound of a string can be distinguished from the sounds of its parts by the coexistence of other sounds with the fundamental sound, and by the softer sound of the divided parts. But one would never be able to distinguish by ear the more or less great numbers of parts into which a string is

---

[1] Descartes suggests that the soul is most satisfied by objects of the senses which are neither too easy, nor too hard, to understand.—*CBH*

divided, if one did not know the attributes of the string. In the same way, in listening to the sound of a plate, as in the experiments explained in Part II, Sect. 7, one will not be able to judge the shape of the plate or the mode of its vibration by listening, except that the sounds of those figures where the interior is surrounded by nodal lines are more sonorous than those of the figures where there are only divergent lines towards the edge.

240. Timbre and Articulation of Sounds

The different timbre of sounds and their articulation are one of the more remarkable purposes of the ear. It doesn't appear to be dependent on the mode of the vibrations, nor the form of the sounding body (or only slightly), but rather (Par. 31) on the material of the sounding body and that of the body with which it is rubbed or hit that propagates the sound. We do not have the least idea of the nature of these different characteristics of the sound, or of their propagation.

241. Distance of a Sound

One does not have the direct sensation of the distance from the point where the sound is produced; but the intensity often serves to determine it, according to certain measures that were developed from the experiments. An increase in the intensity leads to the belief that the object which produced the sound is approaching, and a decrease in the intensity leads to the presumption that it is receding.

242. Direction of a Sound

The best research on the manner in which the direction of sound can be determined by the ear was done by Venturi (*Voigt's Magazin*, vol. 2, ch. 1). If one of the ears is blocked, the eyes are blindfolded, and one stays in the same position, the sound always appears to come from the side of the open ear, no matter where it is placed or how it is produced. The object which produces the sound appears to be on the acoustic axis of the ear. When the intensity of the sound remains the same, and the head is turned successively toward all of the points of the horizon, one hears the sound more or less strongly, depending on whether the acoustic axis of the open ear approaches or recedes from the direction of the sound. One will be able, therefore, to judge direction of the sound by the effect of the sound on an ear. When both ears are opened one will be able to determine the direction by the inequalities of the effect on one and the other, except that when the listener remains stationary, one cannot distinguish between sounds that are produced in front or behind. It appears that animals sometimes turn their ears from side to side to inform themselves of the direction of the sound.

Section 2: The Hearing of Different Animals

243. General Remarks

In comparing the auditory organs of man with those of different animals, one finds that, in all animals, the essential organs necessary for hearing consist of a gelatinous pulp, wrapped in a very fine and elastic membrane, in which the ends of the auditory nerve are resolved. Some other sections, designed to enforce or to modify the sound, are not found in all animals, and their structure varies greatly.

244. Essential Organs Necessary for Hearing

The simplest auditory organs are observed in some crustaceans. In crayfish, one finds a scaly cylinder at the base of the antennae, the substance of which is harder than that of the head. The exterior end of this cylinder is closed by an elastic membrane that Minasi (*De' timpanette dell' udito scoperti nel Granchio Paguro*, Nap., 1775) and Fabricius (*Nov. Act. Hafniens*, 1783) take for the round window. The cavity of this cylinder contains a membranous sac full of gelatinous water, in which is located the substance of the auditory nerve. It enters from the interior end of the cylinder and has the same origin as the nerves of the antennae. Because of the exterior membrane, it appears that these organs are designed to hear as well in air as in water.

Comparetti, who described the organs in detail, also found a small auditory bone in the shape of a nail in the *cancer hastatus*. The bone is wrapped in a membrane. The point of this small bone is pointed inwards.

In several insects, Comparetti thought he also observed little sacs or transparent tubes enveloped in fine membranes, which appeared to be auditory organs; for example, in beetles, grasshoppers, butterflies, moths, hornets, bees, flies, ants, spiders, etc.

© Springer International Publishing Switzerland 2015
E.F.F. Chladni, R.T. Beyer, *Treatise on Acoustics*,
DOI 10.1007/978-3-319-20361-4_17

245. Auditory Organs of Cuttlefish, Octopi, and Squid

Auditory organs are not found in any species of mollusks or worms, except in *cuttlefish (sepia)*, *octopi*, and *squid (loligo)*. They are as simple as those of the crayfish and approach those of fish. In the ring-shaped cartilage that forms the base of the feet or tentacles, there are two irregular oval cavities, separated by a wall. Each of these cavities contains a sac filled with a gelatinous pulp, in which a small body is suspended. The substance of this body is bony in the cuttlefish and resembles starch in the octopus. The sound is only sensed by the motion of the head.

246. Auditory Organs of Fish

Fish with free gills do not have any exterior opening at all; they hear, therefore, only by the motion of the head. Their labyrinth contains three semi-circular canals which pass into a sac. Each of these canals has a bulge in the shape of a bulb near the point where it penetrates the sac. Two of these canals reunite; so that there are only five openings that communicate with the sac, as are found in other higher classes of animals. The bag, near bursting with gelatinous pulp, contains rocks or small bones, of which the number (one to three), the shape, and the hardness vary greatly. They are suspended by a large number of nerve fibers. All of these organs are contained in the same cavity as the brain, and the bones are only found in some recesses.

In *fish with fixed gills*, or *chondro-pterygiens*, such as the rays and dogfish, one finds the same parts as in those preceding, but arranged in a different way. They also have an opening that one could regard as a round window, closed by a thin membrane and covered by ordinary skin. The small bones or rocks that the bag contains are less consistent than the ones in the preceding cases. The entire labyrinth is contained in a special cavity which only communicates with the brain cavity via the holes through which the nerves pass. The auditory organs of these fish seem to be intermediate between those preceding and those of the reptiles.

247. Auditory Organs of Reptiles

In reptiles, the auditory organ is composed of the same parts as those of fish, but some species have one additional part.

Salamanders have three canals and a sac which contains a rock of the consistency of starch. The organ is contained in the cranium, as in the fish with free gills. Their oval window is closed with a small cartilaginous cover.

Snakes have the same parts and an oval window covered with a plate of small bone, whose exterior extremity touches the skin behind the joint of the lower jaw.

The *caecilia americaine* has also a type of eardrum to which this bone transmits, and a Eustachian tube.

Frogs, toads, lizards, and *turtles* have the same parts as fish, but also a cavity, an eardrum (with the exception of the *chameleon* and several other species), a Eustachian tube, a bone, and a trace of a cochlea. The shape and disposition of the parts vary widely.

248. Auditory Organs of Birds

In *birds*, the auditory organs resemble slightly those of the land reptiles, except that they do not have stones, but a less twisted cochlea like that of man or quadrupeds. The oval window is closed by a small bone that connects to the eardrum. There is also a round window, by which the sound is transmitted to the labyrinth in two ways. The cavity leads to three great cavities covered by bony thin blades and an elastic membrane that appear to serve to reinforce the action of the sound on the labyrinth.

249. Auditory Organs of Mammals

In *mammals,* one finds the same auditory organs as in man, as described in the preceding section, but the dimensions, the shape, and the distribution are not the same in all these animals. *Marine mammals* have the same organs as the other mammals, but the cochlea is stronger and slightly elevated. The canals are very thin and the bony blade that forms the cavity is rolled towards itself in the shape of a shell. In general, the labyrinth of the mammals is smaller than that of the birds.

250. Summary of the Organs Found in Different Animals

The organs required for hearing are found, therefore, in all animals examined up until now; but some auxiliary organs, designed to hear more perfectly, are located only in some classes of animals. As for the *labyrinth*, as the principal center of operations of the ear, crayfish and cuttlefish appear to have a *vestibule*, and the organs of some insects, which appear to be designed to hear, are not well enough known to compare them with the auditory organs of other animals. All of the other classes of animals have, besides the vestibule, three canals that expand in bulbous form before rejoining in the sac of the vestibule. Hot-blooded animals have a cochlea, and the others have small bones or suspended rocks in the sac of the vestibule. In most animals, the substance of the auditory nerve appears in two ways; in pulpy form in the canals and in the sac, and in fibrous form elsewhere.

The round window is found in all animals that have a cochlea.

The oval window is found in all animals (except insects and squid); in some animals it is formed by a bony cover of cartilage, and in others by a small bone. The two windows are not always in oval and round shapes. Therefore, it will be more suitable to name them *vestibular window* and *cochlear window*.

A *cavity* and a *Eustachian tube* are not found in all animals that have an eardrum. These are lacking in insects, worms, some snakes, and salamanders. In mammals, it is concave on the outside, and in birds and some reptiles it is convex. The cavity of mammals contains four *small bones*; birds and reptiles have only one. The *auditory meatus* is found only in mammals and birds, and the *external ear* is found only in most mammals.

251. Authors Consulted

The principal authors who have published these observations regarding the auditory organs of the different animals are:

Ant. Scarpa, in *Anatom. Disquisit. de auditu et olfactu,* Ticin, 1789.

And. Comparetti, in *Observat. Anatom. de aure interna comparata,* Patav, 1789.

Leçons d'Anatomie comparée, de G. Cuvier, lesson xiii, *De l'organe de l'ouïe*, contains much new research.

P. Camper has published observations on the auditory organs of fish and cuttlefish, in the papers presented at *l'Acad. de Paris*, vol VII, p. 177, and in the *Memoirs of the Haarlem Society* (*Verhandlingen der Haarlemer Maatschappye*), vol. VII, p. 1, vol. IX, p. 3, and vol. VXVII, p. 2. The papers of Camper are also translated into German (*Kleine Schriften,* vol. I and II), but the work is not found in the three-volume French translation.

John Hunter has described fish in *Philosoph. Transact.*, vol. 72.

The structure and physiology of fishes explained and compared with those of man and other animals by Alex. Monro, Edinburgh, 1785. Chapters VII, IX, and X contain much research on the auditory organs of marine mammals, cuttlefish, fish, and sea turtles. A German translation of this work, by J. G. Schneider with annotation by P. Camper, was published in Leipzig, 1787.

Etienne Louis Geoffroy, *Sur l'organe de l'ouie de l'homme, des reptiles et des poissons*, Amsterdam and Paris, 1778. This memoir can also be found in Vol. II of the Memoirs presented at the Academy of Sciences in Paris; a German translation was published in Leipzig, 1780.

Kohlreuter, in the *Nov. Comment. Acad. Petrop.*, vol. XVII.

Appendix A: Program of the *Institut de France*, in Which a Prize Is Proposed for the Mathematical Theory of Vibrating Plates

Institute de France: Class of the Mathematical and Physical Sciences

Mathematics Prize

The first research on sound dates back to high antiquity; Pythagoras is attributed with the discovery of the ratios between the length of the strings which render different tones; but this field of physical–mathematical science did not develop, and has only made notable progress since the end of the seventeenth century.

It was Sauveur, elected member of the *Paris Academy of Science* in 1696, who has the glory of having developed the theory of vibrating strings and its application to music, one of the important branches of physics, and of having connected it to mechanics. This scholar found, or at least made perceptible by very ingenious methods, the divisions of a sonorous string into several waves separated by nodes or points of repose, which takes place in certain circumstances. To the knowledge of the ratios between the numbers of vibrations and the tones, he added the determination of the absolute number of vibrations which constitutes every tone, concluded from fine and curious experiments, and compared them with analytical formulae which he deduced from the theory of the centers of oscillations (*Mémoires de l'Académie*, 1713).[1]

Taylor, in his *Methodus incrementorum*, published in 1717, addressed the problem in depth, from an analytic point of view, in supposing that the forces that animate the material points of the system are proportional to their distance from a line drawn between the fixed points, and that, consequently, these points arrive at

---

[1] The Report is printed following this Program.—*Note from original text*

© Springer International Publishing Switzerland 2015
E.F.F. Chladni, R.T. Beyer, *Treatise on Acoustics*,
DOI 10.1007/978-3-319-20361-4

this line all together. Twenty or thirty years later, Daniel Bernoulli added many developments to the theory of Taylor, but the general solution and rigorousness of the problem is due to d'Alembert and Euler; these great geometricians were the first ones to employ the differential equation of the motion of the sonorous string, which is in second-order partial derivatives. This equation was first found and integrated by d'Alembert; but Euler had a better sense for the very generality of the integral. One of the geometricians of the Class subsequently published papers on the same subject, where the material is treated with the clarity and depth that characterizes all of his productions.

An equation of the same nature and of the same order as that of the vibrating string applies to the oscillation of air in pipes. The order of the equation does not change when one passes from the linear case (treated first by Lagrange and which Euler seems to have then exhausted) to the cases of two and three dimensions, with which Euler and the other great geometricians are also occupied. Mr. Poisson recently read a very good dissertation on this subject to the Class, which was crowned by his testimony.

In the problems of which we have just spoken, the order of the differential equation of the motion holds in the way that one envisions the effects of the elasticity on the bodies that are animated by this motion. Thus, for example, if it is a question of the sonorous string, which is given a certain tension between two points that have been rendered immobile, the elasticity of this string, that one supposes is without natural stiffness, can take place only in the direction of its length. The effect of this elasticity, when one elongates the string a little by bending, consists of giving it a continual tendency to put itself back in the rectilinear position between the two fixed points. If one assumes that one of these points is immobile and the other made free, the perfect flexible string is no longer capable of producing acoustic phenomena.

Things happen completely differently if the string becomes a *spring*, naturally assuming a certain shape. When all its points are free, it always returns to this same shape, when it has been changed by exterior forces and when the spring no longer has a fixed point.

In this last case, and while limiting oneself, if one wishes, to a single fixed point, the rod or blade that is a spring will return a perceptible sound when it is put into vibration, as long as the number of the oscillations is at least at 25 per second. The differential equation of the motion, which was of the second order in the case of the flexible taut string, is found to be of the fourth order for the *springy* rod; the primary problem can be looked at as a particular case of the second, while disregarding the *spring*, but the inverse does not take place.

This basic difference between the questions of motion, considered under each of these points of view in the simple linear case, shows right away that the same type of differences, and a great increase in difficulty, will be found when introducing two dimensions into the calculation. The acoustic phenomena that are offered by the membranes or taut skins of the drums and kettledrums relate to those of the taut string, without natural stiffness; vibrations of the planes or metallic blades are in the class that includes spring rods.

Euler, in his dissertation *de Motu vibratorio tympanorum*, seeks to relate the vibratory motion of the taut *membranes* to that of the nonrigid string, in considering these membranes as tissues composed of filaments that cross at right angles. One of the geometricians of the Class has published research on this matter in one of our volumes, where he looks at the question from the same point of view; the differential equation of the motion, partial of the second order, cannot be integrated, at least in finite terms.

The same Euler, in his dissertation *de Sono campanarum*, has also tried to relate the vibrations of the rigid surfaces of revolution to those of the rings or circular lines that are springs, in considering these surfaces as assemblies of similarly situated rings in perpendicular planes at the axis of revolution, and while supposing that the effect of vibrations consists of variations in the lengths of their diameters. He arrives at an equation of the partial differential of the fourth order, so that it makes up the nature of the question, which cannot be integrated in finite terms.

This is all that the geometricians were able to do on the problem of sounding bodies, considered in the case of two dimensions. They introduced some simplifications which (it cannot be concealed) change the natural state of things in a way that the results of the analysis are not applicable.

These hypothetical simplifications are especially inadmissible when it is a question of vibrating metallic surfaces, or those enjoying a natural elasticity. Taking the simpler case of the plane, it has been demonstrated that one cannot apply the assumption of Euler to the moving surfaces, reducing the vibrations to simple changes in the curved shapes that one can trace on this plane.

One does not, therefore, get the same differential equations of the motion for this type of vibration, while contemplating their phenomena such as nature gives them. The research of these equations alone would offer to the geometricians a very interesting subject of meditation that can contribute equally to the progress of physics and to that of analysis.

Fortunately, relative to the vibrations of elastic surfaces, we find ourselves in a similar position as the one in which Sauveur put physicians and geometricians at the beginning of the eighteenth century. Mr. Chladni has been occupied for several years with the examination of acoustic phenomena that are offered by elastic blades. He has discovered and rendered perceptible in these blades, in a very ingenious manner, *vibrating layers* analogous to the *waves* of the strings of Sauveur, and the curves of *equilibrium* or of *repose* which correspond to the *nodes* or *points of repose* of the same strings.

His Majesty the Emperor and King, who has deigned to call Mr. Chladni to him and to look at his experiments, struck by the influence that the discovery of a rigorous theory, that would explain all of the phenomena made perceptible by these experiments, would have on the progress of physics and analysis, desires that the Class select the subject of a prize which will be proposed to all of the scholars of Europe. This new conception of the beneficial genius that leads and directs the great and profound views of His Majesty for the progress and propagation of wisdom will be received appreciatively by all people who honor and cultivate the sciences.

Therefore, the Class proposes for the topic of the prize: to give the mathematic theory of the vibrations of elastic surfaces, and to compare it with experiments. The prize will be a gold medal, valued at 3000 francs; it will be awarded in the public session of the first Monday of January 1812. These works will only be received until October 1, 1811; this deadline is firm.

Appendix B: Reports on the *Clavicylinder* and on the Acoustic Research of the Author

Adopted by the Class of Mathematical and Physical Sciences, and by Those of the Fine Arts, on a New Musical Instrument Invented by Mr. Chladni, in the Sessions of December 19 and 24, 1808

Mr. Chladni, correspondent of the Academy of Petersburg and member of several other scholarly societies, presented to the Class of Mathematical and Physical Sciences, and to those of the Fine Arts, a musical instrument of his invention which is called the *clavicylinder*, and a work containing research on the mathematic and physical theory of sound. His instrument was heard, and he explained the principal points of his theory to a commission composed of members taken from the two Classes, who will first give their opinion on the primary subject, and who will next do a special report on the second subject.

The clavicylinder is a keyboard instrument, of nearly the same shape as the pianoforte, but smaller in size. The length is 0.8 m, the width is 0.5 m, and the depth is 0.18 m. The spread of his keyboard is four and a half octaves, from the deepest *do* up to the most acute *fa* of the harpsichord. When one wants to play this instrument, one turns a glass cylinder placed in the box between the inside end of the keys and the backboard of the instrument, by means of a pedal crank provided with a small wheel. This cylinder is the same length as the keyboard and is parallel to it; and when the keys go down, they rub against the surface of the bodies which produce the sounds.

The author makes a secret of the interior mechanism; the sounding bodies are hidden; only the cylinder is visible; and it is presumed that this piece itself would be hidden without the necessity of having to wet it time after time when playing the clavicylinder.

We can therefore only realize the musical effect of the instrument on which Mr. Chladni, who is equally skillful in the theory and the practice of music, played several pieces for us that we heard with the greatest pleasure. This instrument has

© Springer International Publishing Switzerland 2015
E.F.F. Chladni, R.T. Beyer, *Treatise on Acoustics*,
DOI 10.1007/978-3-319-20361-4

many similarities to the harmonica, according to the quality and the timbre of the sound, without exciting a discomfort and an irritation in the nervous system in some very sensitive individuals that puts them in a state of suffering, as the harmonica does.[2]

The clavicylinder also has the advantage over the harmonica in that the graduation of intensity of sounds is better nuanced between the *high-pitched* and the *bass*; in this regard, it is superior to the *bourdon*,[3] which is played in a chamber organ and to which it can be compared.

It was important to know if each of the sounding bodies contained in the box produced the sound promptly as soon as its key was lowered. To reassure ourselves, several of us put a hand on the keyboard and recognized that the clavicylinder left very little to be desired in this regard.

Mr. Chladni assures that the harmony of the instrument is inalterable when its interior parts have been adjusted and regulated, once and for all. We are not penalized in believing it, as much for the confidence which he deserves, as for the plausible conjectures that one can make on the nature of the sounding bodies that it employs. We must also grant that, as is true for all keyboard instruments, the black keys take on the double function of the sharps above and the flats below.

But what essentially distinguishes and characterizes the clavicylinder is the precise correctness that it gives to spun sounds.[4] By pressing the key more or less, one can increase these sounds at will and by the most imperceptible nuances. Above all, it possesses this quality to an eminent degree, from *medium* intensity to *smorzando*.

The limits between this *medium* and the *maximum* of the *rinforzando* are not very great, seeing that the instrument does not have a very forceful sound. If one wants to preserve the beauty of the timbre in all its purity, it is not necessary to press the key too hard; so to use it, in its current state, with the effects of an orchestra, it would be necessary, for larger halls, to gather together several. Nevertheless, we have reason to believe that the clavicylinder can be perfected in this regard, and that, while increasing the interval of *piano* to *forte*, according to the intensity of the sound, one will increase at the same time the difference between the smallest and the greatest pressure of the keys, compatible with the beauty of the execution.

Although we do not know, as we have seen previously, the interior mechanism of the clavicylinder, we are certain that this mechanism is essentially different from those that have been adopted by several other keyboard instruments, set up to obtain continuous sounds, either by metal or catgut strings, by rubbing against the strings with types of bows, chains, or endless loops, etc. One of us heard, in Paris,

---

[2] While the glass harmonica was originally believed to have calming and even healing qualities, because of its hypnotic sweetness, it was eventually thought to cause nervous disorders, animal convulsions, marital disputes, premature childbirth, and mental illness.—*CBH*

[3] Name derived from the French word for buzz; refers to a type of organ pipe.—*MAB*

[4] *Son filé*—the progression of a musical note from weak to strong and then returning back to silence.—*MAB*

approximately 30 years ago, a type of harpsichord that was called an *aéroclavicorde*, whose metal strings were made to resonate by directing currents or streams of air on them, which gave a lively impulse with a very strong bellows. The sounds were of great beauty; but this instrument, totally different from that of Mr. Chladni, did not offer any resources for the *rinforzando* and the *smorzando*. It was also inconveniently slow in the production of sound, which was only heard at the end of a perceptible length of time after the lowering of the key.

The clavicylinder, exempt from this defect, was able to return quick successions of sounds and the trill effect, and also lends itself to the execution of the allegro. But to allow it to produce all of the effects of which it is capable, it is necessary, above all, to apply it to pieces of a tender, melancholic, and even sad nature. Mr. Chladni played for us several of these diverse genres, which have a really charming expression on his instrument, and which made the whole party conceive what a skillful musician could draw from it, to express with truth and energy the feeling that animates him. The successions of chords, the holding of harmony, cold on the organ, and dry on the harpsichord, take on life and color on the clavicylinder, and offer to the composer a means of varying and enriching his tableau.

As Mr. Chladni's project has made his instrument well heard by the public, we dispense with entering into greater detail. His invention appeared to us to add new resources to those who possess the musical art, and merits the approval of the two Classes to which it was presented.

Signed, PRONY, *reporter;*
LACÉPÈDE, HAÜY, *members of the Class of Physical and*
    Mathematical Sciences;
GRÉTRY, GOSSEC, MÉHUL, *members of the Class of Fine Arts;*
JOACHIM LEBRETON, *permanent secretary of the*
    aforementioned Class.

The conclusions presented in this Report have been adopted by the Class of Physical and Mathematical Sciences and by the Class of Fine Arts.

Certified conforming to the original.

The permanent secretary for the Mathematical Sciences, signed DELAMBRE.

Adopted by the Class of Mathematical and Physical Sciences, and by Those of the Fine Arts, in the Sessions of February 13 and of March 18, 1809, on the Work of Mr. Chladni, Relative to the Theory of Sound

The Classes of the Sciences and of the Fine Arts heard a report, on the 19th and 24th of last December, on a new musical instrument invented by Mr. Chladni, which must be followed by another report on research presented to these two classes by the same author, relative to the theory of sound. The joint committee introduces the results of the examination made of this research. Mr. Chladni, who has devoted a considerable amount of time to these experiments on sounding bodies, employed very usefully for the progress of science, published his results in a 1787 dissertation containing interesting discoveries on the physical theory of sound.[5]

One section of that dissertation treats the vibrations of rods, both rectilinear and curved, and the sounds that are obtained. Another section, which is of particular interest to physicists, contained new and very curious facts on the vibrations of elastic surfaces. Our colleague, Mr. Haüy, after having learned of it, has repeated experiments in front of the members of the Philomatic Society, using the means by which Mr. Chladni makes visible the division of a vibrating surface into several partial *layers*, each having their distinct oscillations, that correspond to those of the *waves* of a sounding string. These *layers* separated one from another by the curves of equilibrium which represent the *nodes* or *stationary points* of the same string; the *waves* and the *nodes* of the sounding string, as it behaves here, were discovered, or at least made perceptible, by Sauveur more than a century ago.

At the end of this [1787] work, the author promised further extensive details on the material that was his subject, and he kept his promise in publishing a second treatise on this same material that contains everything important from the first one, with considerable additions. This treatise, published in 1802, is written in German, and Mr. Chladni, who suggested making a French translation during his stay in Paris, wished to subject it to the judgment of the Institute before making it public.

The work, under the title of *d'Acoustique* [*On Acoustics*], is divided into four parts that address, respectively:

1. Numeric ratios of the vibrations of sounding bodies.
2. Laws of the phenomena that they offer.
3. Laws of the propagation of the sound.
4. The physiologic part of acoustics, where the author examines the concerns of the sensation of sound and the hearing organs in men and animals.

In general, the first part, which addresses the numeric ratios of the vibrations of the sounding bodies, only contains things that are known. The author proposes, as Sauveur had done in 1713, to adjust the tones of the keyboard, relative to the absolute number of the vibrations, in such a way that the first *do* produces a number of vibrations equal to 128 or to the seventh power of 2, by means of which the different octaves of this fundamental sound also respond to the full force of the same number 2. Knowing the ingenious procedure imagined by Sauveur for determining the absolute number of vibrations given by one of the tones of the

---

[5] In 1787, Chladni published *Entdeckungen über die Theorie des Klanges* (*Discovery of the Theory of Pitch*), in which he first outlined his experiments with sand, glass plates, and a violin bow. *Die Akustik* (*On Acoustics*) was published in 1802, in German, and *d'Acoustique* (a revised and expanded version and the subject of this translation) was published in 1809, shortly after the material was introduced in this lecture series.—*MAB*

musical scale,[6] Mr. Chladni uses another which consists of causing a strip of metal, fixed at one end, to vibrate for rather a long time, so that one can count the oscillations or vibrations that it makes during a set length of time. Their number will be the same as those of another blade, during the same period of time and in the same circumstances, in reverse ratio to the length of the blade.

In this first part, Mr. Chladni also treats *temperaments*[7] proposed by various people. He gives preference to that which was adopted by Rameau, and which rendered the 12 semi-tones contained within the limits of an octave perfectly equal between them, and matching them to 12 geometric means taken between the extreme terms. Some musicians found that this temperament satisfied the spirit more than the ear. In their opinion, the thirds are a little too strong, sacrificed to less altered fifths, although likely a more bearable alteration; but this is not the place to examine this question.

The second part, which treats the laws of the phenomena offered by the vibrations of bodies, is the one where, along with things already known on this subject, one finds the new discoveries of the author which makes this part of his work most original and curious, and worthy of the interest and attention of physicists and

---

[6] When two organ pipes approaching unison resonate together, there is a certain moment where the joint sound that they render is stronger, and these moments seem to return at equal intervals of time. Sauveur supposed, with a lot of credibility, that these swelling sounds, called *beats* by the organists, took place when vibrations, after a certain time of non-coincidence, came together to hit the ear at the same time. According to this ingenious insight, knowing the *interval* between the tones of the pipes (by which the ratio between the numbers of their vibrations is deduced) and the time that passes between two *beats*, the absolute numbers of vibrations of the pipes during this time become the terms of the ratio between the numbers of vibrations, with this ratio being reduced to its simplest expression.

In this way, Sauveur found that an open organ pipe, five feet in length, gave 102 pulsations per second. This pipe is at unison with *la*, understood in the rising range of the deepest *do* of a harpsichord, and it was concluded that the *do* below this, that is, the deepest *do* of the harpsichord, must give 61 pulsations per second.

These experiments were conducted in 1700. Twelve years later, Sauveur compared their results with formulas that he had deduced from the theory of the centers of oscillations, and these expressed the relationship between the time and the number of vibrations of the strings, when one had the necessary data. He was surprised to find, by these formulas, a number of vibrations double those that were deduced from the experiments; but he noticed very quickly that one must distinguish, according to the effect on the ear, the oscillations of a cylinder of air contained in a pipe that produce sensible *beats*, from those that seem to escape the ear and give only insensible beats. He saw, according to this insight, that in his experiments on pipes, he had counted the *comings* and *goings* as a single vibration, instead of the relative calculations of the strings; the *coming* was taken for one vibration and the *going* for another, as and when it behaves like the oscillations of a clock.

Sauveur was determined to take the vibrations as the pipes gave them to him, that is, in counting one *coming* and one *going* as a single vibration that he called *acoustic vibration*. Since 61, the number of acoustic vibrations per second of the deepest *do* of the harpsichord, is little different from 64, which is the sixth power of 2, he raised it a little and assigned the number 64 to this *do*. This is equivalent to the number 128 adopted by Mr. Chladni, in counting each acoustic vibration as two ordinary vibrations.—*Class reporter's note*

[7] A *temperament* is a system of tuning for musical instruments. See Pars. 24–28.—*MAB*

geometricians. He first examines the vibrations of strings and rods, and distinguishes three kinds, namely: *transverse, longitudinal,* and the one that he calls *torsional*. The first ones are those that take place when a string or rod is touched in a direction perpendicular to its length. They are related to the phenomena that, in the last century, were submitted to the analysis of several geometricians, one of whom is member of this Class.

But a rod that returns a certain sound, when touched in this way, will be heard completely differently if it is rubbed in the direction of its length with a piece of cloth, which must be wet for glass, and dry for other bodies. This is already an important class of phenomena in which it appears that Mr. Chladni is the first to engage. He finds that these vibrations in a solid rod, that he calls longitudinal, are subject to the same laws as the longitudinal vibrations of the air in an organ pipe, and gives a table of the frequencies of these vibrations for different materials, such as glass, metal, and wood.

Still different sounds than those produced in the two preceding circumstances are obtained when one rubs a rod in a sharply angled direction from its axis. Mr. Chladni uses the term *torsional* for the vibrations resulting from this type of friction because it is assumed that the molecules of the body move in rotation or oscillation around its longitudinal axis. He says that he recognizes that in these vibrations the numeric ratios are the same as those of the longitudinal vibrations, but that the tones of each rod are lowered by a fifth. It does not appear that others did these experiments before him.

Each series of experiments of which we have just spoken was done on rods that were fixed, simply supported at one or both ends, fixed at one end and supported at the other, or, finally, with both ends free. Each of these circumstances offers specific results. Mr. Chladni also examines the vibrations of curved rods, forks, and rings. Euler wanted to apply this last type of vibration to the phenomena of the sounds of bells, but Mr. Chladni finds, with reason, that his hypothesis does not conform to nature.

The last two sections of this second part are dedicated to the vibrations of plates and bells, or, in general, planar or curved surfaces, a completely new subject in experimental physics, and which, despite the striking and remarkable regularity of the phenomena, has resisted the efforts of skillful geometricians who wanted to address it.

Mr. Chladni has determined the places occupied in the musical scale by the sounds that one can draw from plates in looking at their different shapes, and in making them sound in different ways. But the interest inspired by this research increases significantly when we combine it with the research whose object is the determination of the portions of the surfaces of every plate that have distinct and coexisting vibrations and remarkable curves that serve as boundaries. Mr. Chladni has imagined a simple and ingenious means of rendering these curves visible to the eye. He covers with sand the plate that he wants to resonate. When the sound is produced, the sand abandons all of the oscillating parts of the body, and, remaining stationary on their borders, takes refuge where the curved axes of equilibrium are located, which assume a variety of different, but perfectly regular, shapes.

To perform the experiments, it is necessary to squeeze the plate with the tips of two fingers at two opposite points on its faces, and to rub it with a bow at a point on the perimeter. Sometimes, a third finger is applied at different points of one of the sides to vary the results of the experiments. Instead of holding the plate between the fingers, one of its faces can be set down on a fixed point, and the other face supported at a second point exactly opposite from the first. This is how Mr. Paradisi, of Milan, did these experiments, which we will speak of soon.

The support point is always part of one of the curves of equilibrium; their shapes and the arrangement of their composition depends on the shape of the plate, the position of the point held, the position of the point where one applies the bow, and, finally, on the different sounds that one wants to obtain while rubbing the bow in different ways at the same point. If one or more of these circumstances change, the shapes of the bends and the arrangement of their composition also change.

In reporting on these curious phenomena, we cannot avoid speaking about a dissertation that contains its own research to establish the systems and the connections between them, and which is included in the first volume of the collection of the Institute of the Sciences of the Kingdom of Italy,[8] under the title of *Ricerche sopra lavibrazione delle lamines elastiche*.

The author of this dissertation is Mr. Paradisi, member of the Institute, and state counselor and director general of public works for the Kingdom of Italy. He states in a note, that he undertook this work after reading a passage in the *Bibliothèque brittanique* about Mr. Chladni's experiments and his way of making the curves of equilibrium visible, in spreading sand on the plates. Provided with a device by means of which he could keep the plates at the fixed points arbitrarily situated on their surfaces, without a helping hand, he recognized at first that the curves of equilibrium only achieved a constant shape after a gradual progression, and continued with variable shapes, the generation of which he has examined with care, driving him to new conclusions on the theory of these curves.

So, for example, if a glass plate in the shape of a parallelogram rectangle, 9 in. long and 3 in. wide, is held on the long axis at one sixth of the distance between the two ends, and a bow is applied against one of the long sides of the parallelogram, at a third of the distance between the ends of this side, the sand lines attain a fixed state, dividing the surface of the plate into eight equal squares, with a line in the direction of the long axis, and three equidistant lines parallel to the short edge. This is the first of Mr. Paradisi's experiments; but he recognized that in making the plate vibrate by a series of small successive strokes of the bow, eight semicircles are obtained first, with their centers and their diameters placed symmetrically on the long sides of the parallelogram, and the point of application of the bow is at one of these centers. These circles (those that rest against the same side) increase gradually

---

[8] The *Istituto Nazionale* was founded in 1802 and consisted of three sections: Physical and Mathematical Sciences, Political and Moral Sciences, and Literature and Fine Arts. The first 30 members were chosen by Napoleon Bonaparte. In 1810, it was reorganized as the *Istituto Reale di Scienze, Lettere e Arti.—CBH*

in size. From separated (at first) they become tangential, and then penetrating, while leaving the perpendicular rectilinear tracks between them on the long side. At the same time that these tracks increase in length, the arcs become flattened as they approach the long axis of the parallelogram, with which they eventually merge.

In other experiments, Mr. Paradisi obtains these complete circles, initially formed on the surface of the plate, and semicircles against the long and short sides of the parallelogram. The velocity of the grains of sand located on the borders diminishes by the same measure that the rays increase.

He calls the center of the circle, that forms around the point of application of the bow, the *center of vibration*, and those of other circles located on the plates are called *secondary centers*, assuming that when the system of curves has reached a fixed state, any element of these curves is influenced by the direct result of several forces, including the actions emanating from the various centers of vibrations. The functions of their distances from the curved element with which they are concerned, requires a differential equation between the coordinates of this element, whose integration demands that one know the form and function that represent the laws of the actions of the forces. He has announced another dissertation about the research on this object.

We must look at Mr. Paradisi's dissertation for the details of his other experiments which are distinguished by interesting changes in position of the support points and the point of application of the bow, but which don't produce anything new in the shapes and the arrangement of the system of curves.

Mr. Chladni ends the second part with the consideration of the vibrations of bells and of curved surfaces in general and on the coexistence of vibrations in the sounding bodies. He speaks of the theory and the hypotheses of Euler on the sound of bells, the musical system of Tartini, supported by experiments which, according to Mr. Chladni, were known in Germany for a long time before Tartini made use of them, and that one can look at as the inverse of those of Rameau. Finally, he treats the combination, which takes place in certain circumstances, of vibratory motion with other types of motion.

In the third part, which has for its subject the propagation of sound, the author first considers this propagation operating in the air and in different gases, and he next examines the case where it take place through the intermediary of liquid and solid bodies. In this part of the work, one will note experiments on the vibrations of various types of gas that the author made with Prof. Jacquin of Vienna, conjectures on the cause of the difference between the theoretical speed and the observed speed of the propagation of sound in air, etc., research on the transmission of the sound in solid bodies, etc.

We can dispense with giving a detailed analysis of his work on these matters, already enriched by the research of physicists and geometricians, and we will be satisfied to indicate the subject of the fourth part of the *Treatise on Acoustics*, under the viewpoint which interests physiology, and that has to be judged by the anatomists.

We seem to be united in giving credit that the discoveries with which Mr. Chladni has enriched the physics of sound could not be more curious and interesting, and that they offer the advantage of presenting new and important phenomena to physicists and geometricians that uniquely excite their curiosity and their competition for finding the explanations and determining the laws. Opening this field to scholarly research will not be the only debt that they will have to the author of the new *Acoustics*. It is rather remarkable that a branch of the natural sciences, where there are still so many fine but difficult problems to address, is the first where the history of the human spirit points out prominent truths, and, which is worthy of attention, founded on a rigorous application of calculation to observation. Everyone knows that the discovery of the ratios between the vibrations of sounding bodies that produce different sounds, attributed to Pythagoras, dates back to the greatest antiquity. These ratios, to speak of it in passing, have been used since ancient times as the basis of a musical system convention that distinguished scholars look at as having been common to the Greeks, the Egyptians, the Chinese, etc. One of the principal properties of this system, in which all of the sounds were generated by the *triple progression,* giving a series of fifths, was to have only one type of tone and one type of semi-tone, and to produce a *call* or an energy *attraction* between the sounds proceeding from the last interval.

It was believed that in this system was found the real generation of the natural melody of man when he sings or plays an instrument with free sounds, without accompaniment. We objected to those who wanted to link it to our harmonic system, since the thirds which it gave, considered as dissonant by the ancients, could not be admitted to the harmony, as they were appreciably stronger than those given by the resonance of the sounding body. Aristoxenus and Ptolemy had already introduced this into the ancient system, by modifying that of Pythagoras. Doubts on the principal unity of our musical system were born out of this discussion.

The theorems on the ratios of the vibrations of sounding bodies have been, until Newton, the only truths well recorded in this part of physics. Since the impetus given to the physical–mathematical sciences by this immortal genius, several geometricians of the first order have occupied themselves with problems related to sound. However, the general complete solution of the least difficult one of these problems, in which a simple wire is stretched between two fixed points, escaped all the resources of integral calculus, in an epoch where this calculus was already enriched with the brilliant discoveries of Newton, Leibniz, Bernoulli, Euler, and d'Alembert themselves, and only particular solutions were able to be obtained by their means. The general solution was nevertheless given first by d'Alembert and soon after by Euler; but a new method of analysis was necessary to get there, which is now one of the great instruments of geometricians in the application of calculus to the phenomena of nature, and must be applied to the physics of sound, which has been the subject of one of its first applications, a distinguished rank in the annals of the human spirit.[9]

---

[9] Before the integration of the equation of the vibrations of strings, Euler and d'Alembert integrated the equations in *partial differentials*; the first, in a paper published among those of the Academy of Petersburg, in 1794; the second in his *Traité de la cause des vents.—Class reporter's note*

D'Alembert and Euler together had some rather long disputes on their respective solutions, and Euler had on his opposing side the advantage of a better sense than d'Alembert for the full extent of the meaning of the *arbitrary functions* that complement the integrals of partial differential equations.[10]

Our colleague, Mr. de Lagrange, has given detailed explanations on everything that has to do with the problem of the vibrating string. These explanations, which leave nothing to be desired, can be found in two good, strong papers published among those of the Academy of Turin.

The problems relative to the propagation of sound have also been the subject of research by several scholars of great merit. We have the great pleasure of recalling one of them to the Class, a young geometrician, who has rewarded, by his approval, the discoveries made in an age where commonly we consider ourselves very happy to be able to understand and appreciate the discoveries of others. These problems on the propagation of sound are the same as those that concern vibrating strings, with the advantage of having been treated with the consideration of two and three dimensions of space.

Nevertheless, Euler wanted to extend the applications of the method of analysis to the vibratory motion of sounding bodies, the method of analysis that had succeeded so well for him for the vibrating string, and on which he was the first to give a Treatise *ex professo* (third volume of his *Integral Calculus*), and published two papers (*New Commentaries of the Academy of Petersburg*, vol. X, 1764); one on the vibratory motion of the kettledrums, and the other on the sound of bells; but physics has not withdrawn the fruit of this work where the deep analytical science of the author shines. In the fourth volume of our Memoirs, our colleague Mr. Biot,

---

[10] Mr. de Laplace demonstrated the *discontinuity of these arbitrary functions*, in his dissertation *Sur les fonctions génératrices*, Academy of Sciences, 1779.

Mr. Monge made this discontinuity very perceptible, and one might say evident, by geometric considerations. The geometricians' opinion on this question of analysis are explained in detail in a paper commemorating Mr. Arbogast, *On the nature of arbitrary functions*, etc. that won the prize proposed in 1787 at the Academy of Petersburg where this paper was printed in 1791.

Note: The rest of this note was communicated to the reporter by one of the members of the Class.

The dissertation, *On the nature of the propagation of sound*, included in the first volume of the Society of Turin, which appeared in 1759, contains the first rigorous demonstration on the discontinuity of arbitrary functions, which were then the subject of the dispute between Euler and d'Alembert. This is demonstrated by the unique application made to the solution of the problem of the oscillations of a taut wire loaded with any number of weights; the case where the number of weights approaches infinity, with the wire or string having a uniform thickness and the same manner of construction, results in the consideration of discontinuous functions. This paper also contains the first rigorous general theory of the oscillations of air in open and closed flutes, and the propagation of sound and echoes in a physical line of air; material that Euler treated exhaustively in the papers of Berlin and of Petersburg. Mr. Poisson has looked to extend the theory to the case of three dimensions.

The equation for vibrating surfaces, given by Mr. Biot in the fourth volume of our Memoirs, is only the equation for vibrating surfaces stretched like drums and kettledrums, and not for vibrating elastic surfaces. This one has not yet been given, and appears subject to some difficulties.—*Class reporter's note*

and Mr. Brisson, bridge and road engineer, took up the question of the motion of vibrating elastic surfaces, considering the elasticity only one flat dimension. This manner of looking at the question (explained in greater detail in the Program cited here)[11]: the details, the analytic results to which one aspires, are not applicable to the problems involved in the experiments of Mr. Chladni.

Besides, the principal object of Mr. Biot was to give an example of his methods, to use the general integrals in infinite terms of the partial differential equations in the resolution of these problems, when these integrals cannot be explained otherwise. According to this purely analytical goal, he did not give any special attention to the physics part of the question and to the phenomena found by Mr. Chladni that were known at the time he published his dissertation.

Euler's dissertation, *De sono Campanarum*, gave rise to observations of the same kind. (See the Program cited here.)

It is necessary to acknowledge that the problem of the vibratory motion of sounding bodies deserves all the more to be attacked by geometricians with all new efforts. Far from being resolved when we consider it with two dimensions, the problem offers still more of the same difficulties in the linear case, when one adds a condition to its fundamental terms. For example, if we assume that the string is the largest variable, the increase in difficulties in the case of two dimensions is easily conceived, when one considers that this case reproduces an infinite number of times the already infinite number of circumstances that complicate the linear case.

We think, according to the details which we have just written, that the two Classes owe distinguished praise to the discoveries of Mr. Chladni, relative to the physics of sound, and that it would be important to direct the attention and the emulation of scholars on the physical–mathematical research to which these discoveries can give rise.

Minutes signed by, DE LACÉPÈDE, HAÜY, MÉHUL, GOSSEC, GRÉTRY, LE BRETON, DE PRONY, *reporters;*

This report and its conclusions have been adopted by the Class of Mathematical and Physical Sciences and by the Class of Fine Arts.

Certified conforming to the original:

The permanent secretary for the Mathematical Sciences, signed DELAMBRE.

The permanent secretary for the Fine Arts, signed Joachim LE BRETON.

---

[11] Program of the *Prix de Mathematiques*—included in Appendix A.—*MAB*

Appendix C: Figures

© Springer International Publishing Switzerland 2015
E.F.F. Chladni, R.T. Beyer, *Treatise on Acoustics*,
DOI 10.1007/978-3-319-20361-4

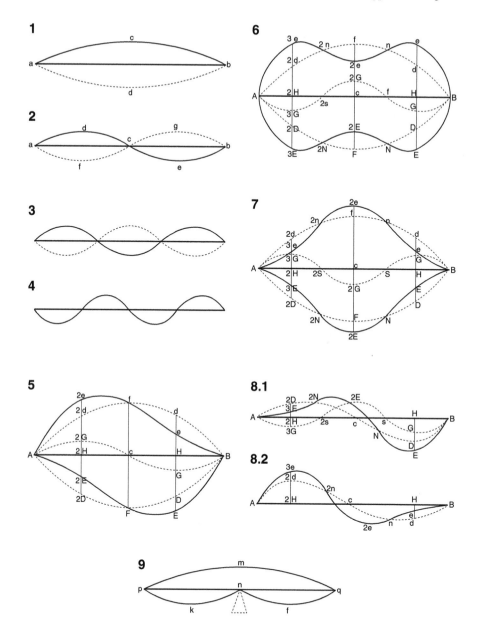

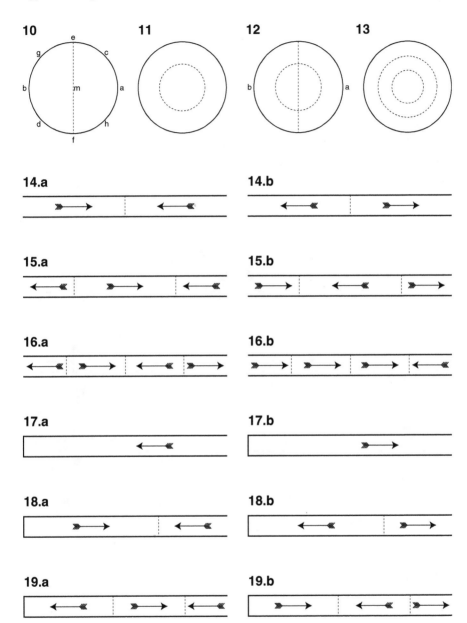

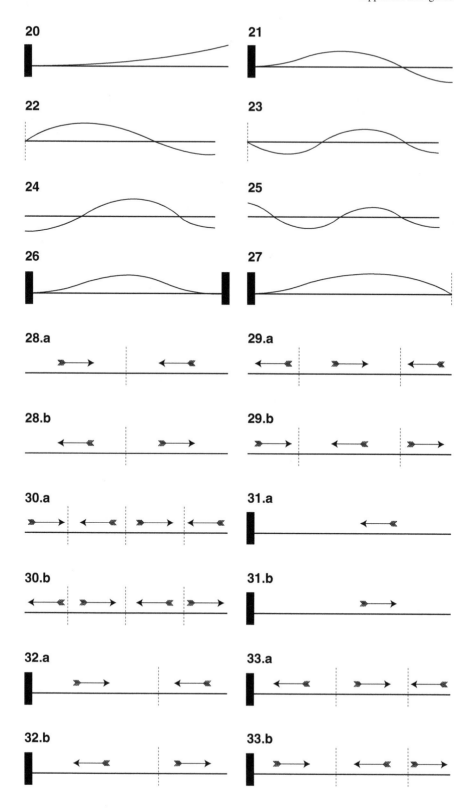

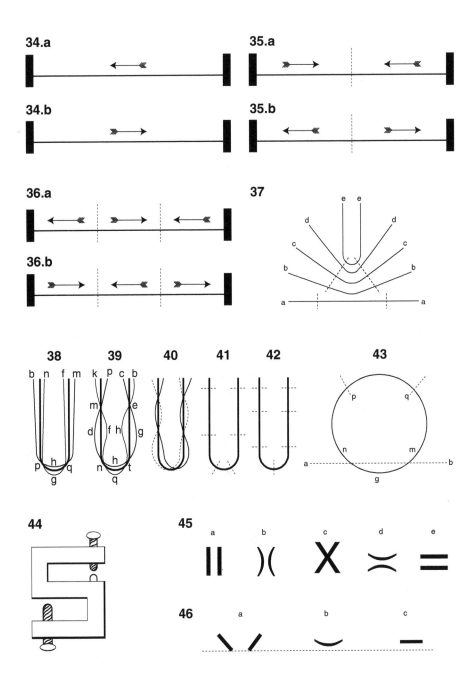

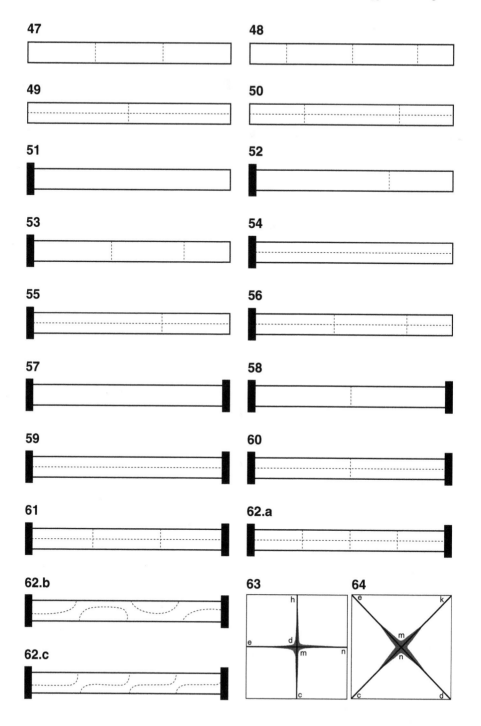

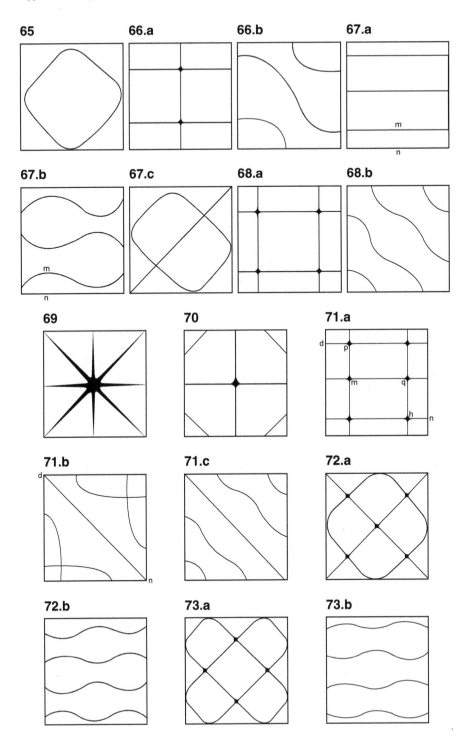

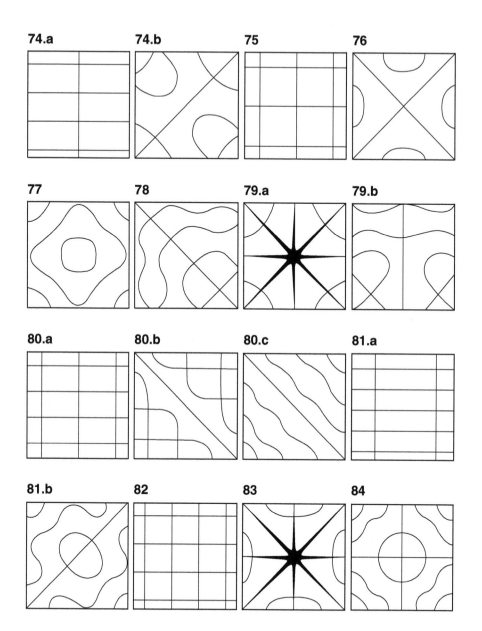

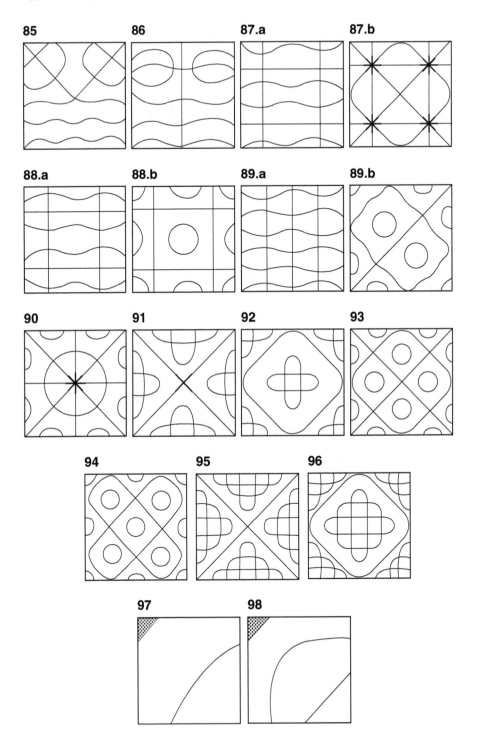

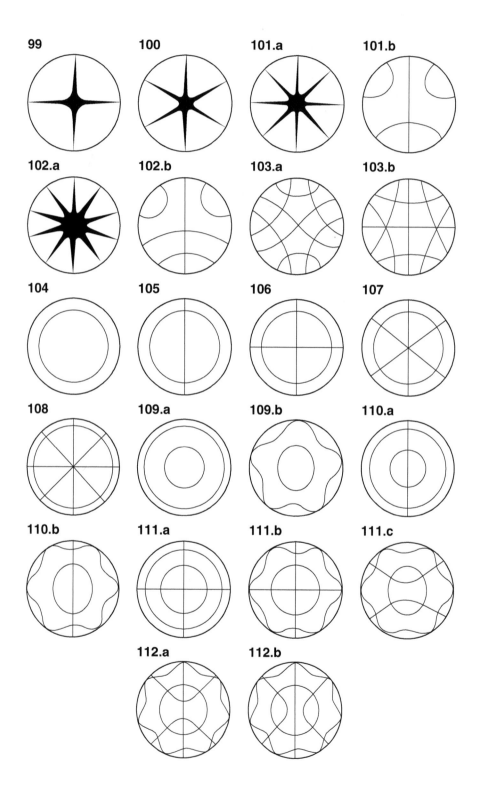

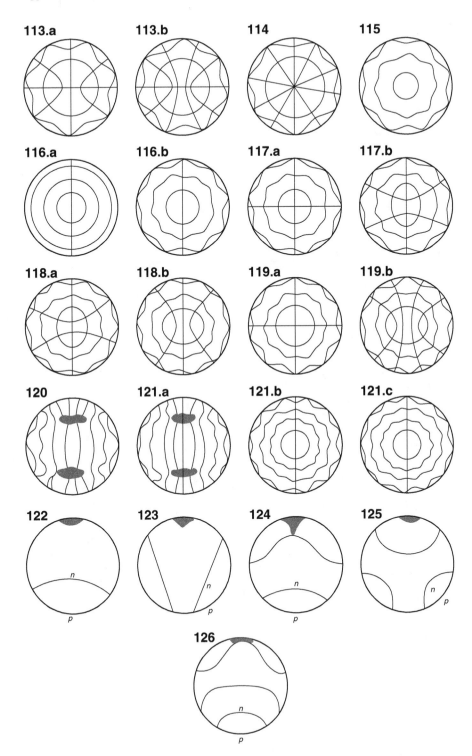

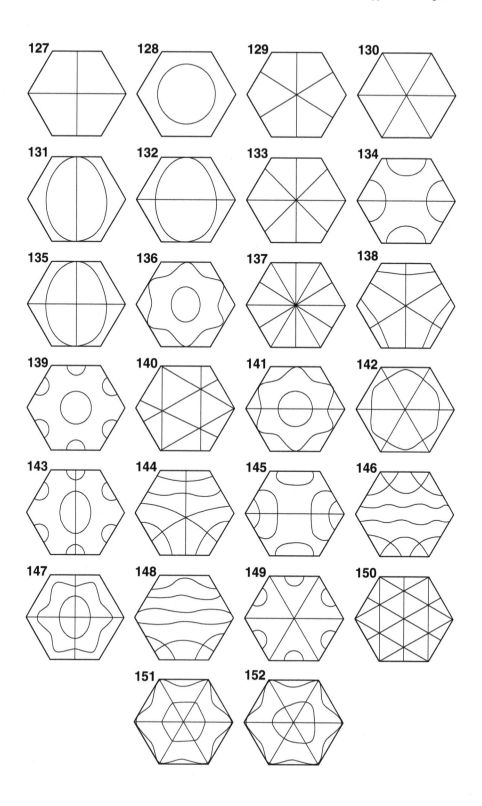

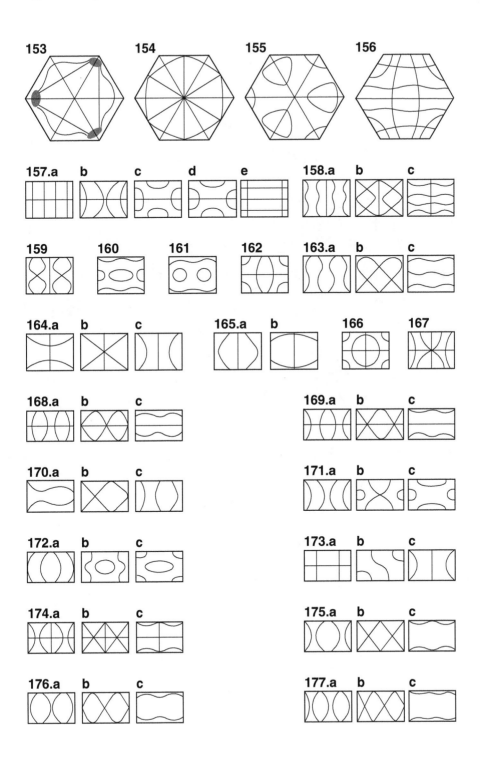

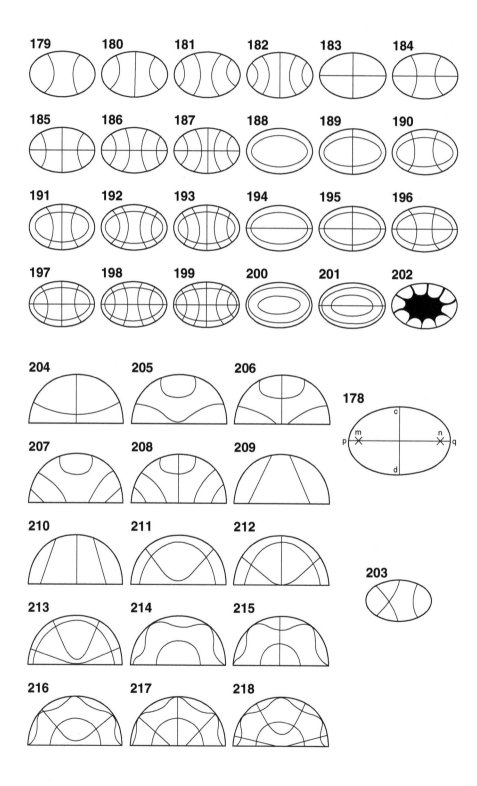

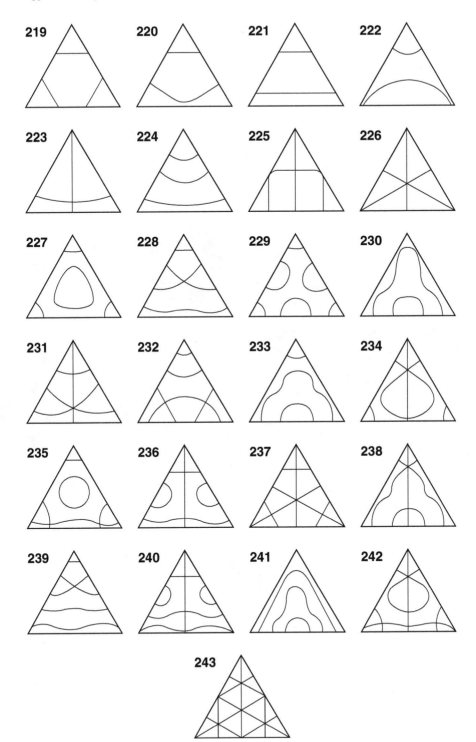

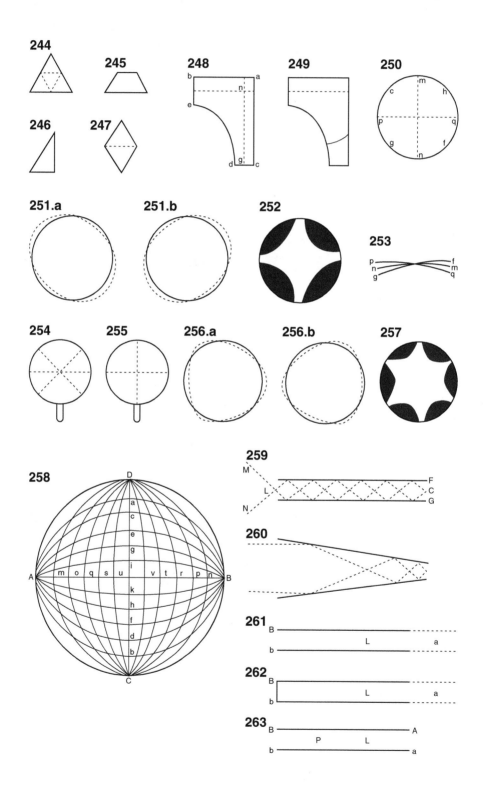

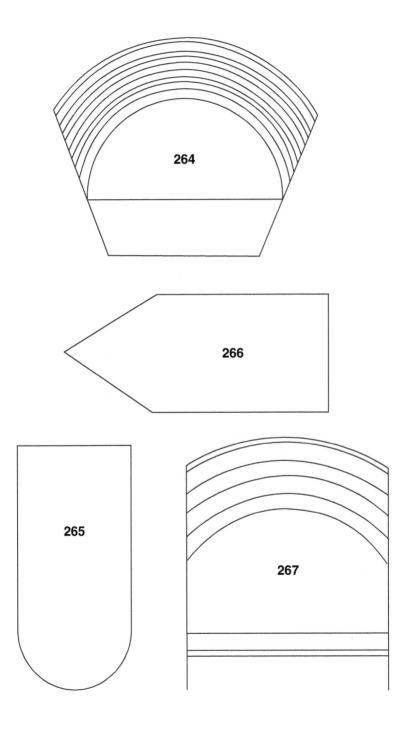

Name Index

Note: Numbers in boldface in this index refer to Chladni's numbered paragraphs and letters in boldface (A and B) are Appendix references. Numbers in regular type are page numbers.

© Springer International Publishing Switzerland 2015
E.F.F. Chladni, R.T. Beyer, *Treatise on Acoustics*,
DOI 10.1007/978-3-319-20361-4

Subject Index

Note: Numbers in boldface in this index refer to Chladni's numbered paragraphs and letters in boldface (A and B) are Appendix references. Numbers in regular type are page numbers.

© Springer International Publishing Switzerland 2015
E.F.F. Chladni, R.T. Beyer, *Treatise on Acoustics*,
DOI 10.1007/978-3-319-20361-4